Rotating Electrical Machines and Power Systems

DALE R. PATRICK

STEPHEN W. FARDO

Eastern Kentucky University

PRENTICE-HALL, INC., Englewood Cliffs, New Jersey 07632

Library of Congress Cataloging in Publication Data

Patrick, Dale R.
  Rotating electrical machines and power systems.

  Includes index.
  1. Electric machinery.   2. Electric transformers.
I. Fardo, Stephen W.   II. Title.
TK2000.P35   1985   621.31′042   84-8241
ISBN 0-13-783309-1

*Editorial/production supervision and
  interior design:* Ellen Denning
Cover design: Whitman Studio, Inc.
Manufacturing buyer: Gordon Osbourne
Page layout: Meryl Poweski

Cover photograph courtesy of Marathon Electric.

Printed in the United States of America

10 9 8 7 6 5 4 3 2 1

ISBN 0-13-783309-1 01

PRENTICE-HALL INTERNATIONAL, INC., *London*
PRENTICE-HALL OF AUSTRALIA PTY. LIMITED, *Sydney*
EDITORA PRENTICE-HALL DO BRASIL, LTDA., *Rio de Janeiro*
PRENTICE-HALL CANADA INC., *Toronto*
PRENTICE-HALL OF INDIA PRIVATE LIMITED, *New Delhi*
PRENTICE-HALL OF JAPAN, INC., *Tokyo*
PRENTICE-HALL OF SOUTHEAST ASIA PTE. LTD., *Singapore*
WHITEHALL BOOKS LIMITED, *Wellington, New Zealand*

Contents

Preface

Rotating Electrical Machines and Power Systems is an up-to-date text book intended for use in technical programs in electrical technology at vocational-technical schools, industrial training programs, and college and university technology departments. The text uses a "systems" format to teach rotating electrical machinery and power system concepts. Key concepts are presented by stressing applications-oriented theory. Through this approach, the student is not burdened with an abundance of information needed only for engineering design of machines. "Real-world" applications and operations are stressed throughout the book. Mathematical problems are solved by basic algebraic and trigonometric operations.

There are very few texts on the market dealing with rotating machines and transformers which are applications-oriented. Several texts are available which deal with engineering design of machines. However, the two-year and four-year college and university market has been somewhat neglected. There seems to be a tremendous demand for an applications-oriented textbook dealing with rotating machines and transformers from a user's or technician's point of view.

The book's organization is very logical and up to date. Specialized machines, such as servo systems, are discussed in detail due to their present use with microprocessor-controlled and numerically controlled machines. Most texts do not thoroughly cover syncho-servo systems or dc stepper motors. The authors feel that coverage of these topics is essential for today's technicians.

Key concepts are presented in the book through an electrical power systems model (production, distribution, conversion, and control systems). The subsystems for producing, distributing, converting, and controlling electrical power are explored in detail. Through this approach students are better able to formulate their understanding of electrical machines and power systems. This approach is different from that of similar texts. The authors have used this instructional method in teaching classes dealing with rotating electrical machines and power systems for many years in a large university technical program.

Much of the artwork from this book has been made available through a number of organizations and manufacturing concerns. In general, this courtesy has been expressed with respective photographs and drawings. In addition to this, some artwork was also made available from the publications *Industrial Electrical Systems* and *Electrical Power Systems Technology* by the authors.

The chapters of the book include an introductory section, the main text, study questions, and problems. Many drawings, photographs, and sample problems are used in the chapters to illustrate key ideas, show important applications, and aid in students' comprehension of material.

STEPHEN W. FARDO
DALE R. PATRICK

Eastern Kentucky University
Richmond, Kentucky

ONE

Introduction to Rotating Electrical Machines and Power Systems

The operation of rotating electrical machines and power systems is rapidly becoming more important. The emphasis on energy conservation and automated manufacturing coupled with the high cost of energy has caused concerns in the areas of electrical power production, distribution, conversion, and control. These areas of electrical technology rely on efficient operation of rotating machines and power systems.

Electrical machines have changed considerably in recent years. The trend is now toward the use of smaller machines and more precision control. Machinery efficiency and design has also become more significant. Electrical machines and power systems have many common characteristics. Among these are operational and construction features. This chapter deals with some of the basic operational considerations of electrical machines and reviews magnetic and electromagnetic principles. Rotating machines and power systems rely on magnetic and electromagnetic circuit action to accomplish their basic operation. In addition to basic operational principles, this chapter also discusses units of measurement which are common for rotating machines and power systems, and presents a model of electrical power systems which is used in the organization of the book.

BRIEF HISTORY OF ELECTRICAL MACHINES

Electrical machinery has been in existence for many years. The applications of electrical machines have expanded rapidly since their first use many years ago. At the present time, applications continue to increase at a rapid rate.

Thomas Edison is given credit for developing the concept of widespread generation and distribution of electrical power. He performed developmental work on direct-current (dc) generators which were driven by steam engines. Edison's work with electrical lights and power production led the way to the development of dc motors and associated control equipment.

Most early discoveries related to electrical machinery operation dealt with direct-current systems. Alternating-current (ac) power generation and distribution became widespread a short time later. The primary reason for converting to ac power production and distribution was that transformers could be used to increase ac voltage levels for long-distance distribution of electrical power. Thus the discovery of transformers allowed the conversion of power production and distribution systems from dc to ac systems. Presently, almost all electrical power systems produce and distribute three-phase alternating current. Transformers allow the voltage produced by an ac generator to be increased while decreasing the current level by a corresponding amount. This allows long-distance distribution at a reduced current level, reduces power losses, and increases system efficiency.

The use of electrical motors has increased for home appliances and industrial and commercial applications for driving machines and sophisticated equipment. Many machines and automated industrial equipment now require precise control. Thus motor design and complexity has changed since early dc motors which were used primarily with railroad trains. Motor control methods have now become more critical to the efficient and effective operation of machines and equipment. Such innovations as servo control systems and industrial robots have led to new developments in motor design.

Our complex system of transportation has also had an impact on the use of electrical machines. Automobiles and other means of ground transportation use electrical motors for starting and generators for their battery-charging systems. There has recently been emphasis in the development of electric motor-driven automobiles. Aircraft use electrical machines in ways similar to automobiles. However, they also use sophisticated synchro and servo-controlled machines while in operation.

REVIEW OF MAGNETISM AND ELECTROMAGNETIC PRINCIPLES

Magnetism and electromagnetic principles are the basis of operation of rotating electrical machines and power systems. For this reason, it is necessary to review basic magnetic and electromagnetic principles.

Permanent magnets.   Magnets are made of iron, cobalt, or nickel materials, usually in an alloy combination. The ends of a magnet are called north and south poles. When a magnet is suspended in air so that it can turn freely, one pole will point to the north magnetic pole of the earth, since the earth is like a large permanent magnet. The north pole of a magnet will *attract* the south pole of another magnet. A north pole *repels* another north pole and a south pole *repels* another south pole. The two laws of magnetism are: (1) like poles repel; and (2) unlike poles attract.

The magnetic field patterns when two permanent magnets are placed end to end are shown in Figure 1-1. When the magnets are farther apart, a smaller force of attraction or repulsion exists. A *magnetic field,* made up of *lines of force* or *magnetic flux,* is set up around any magnetic material. These magnetic flux lines are invisible but have a definite direction from the magnet's north to south pole along the outside of the magnet. When magnetic flux lines are close together, the magnetic field is stronger than when they are farther apart. These basic principles of magnetism are extremely important for the operation of electrical machines.

Figure 1-1 Magnetic field patterns when magnets are placed end to end.

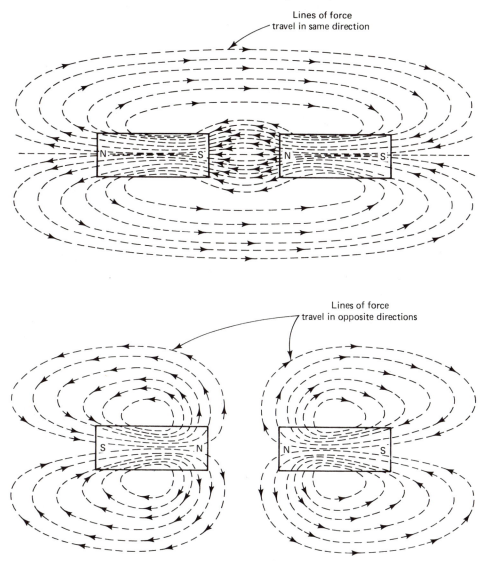

Magnetic field around conductors.   Current-carrying conductors, such as those in electrical machines, produce a magnetic field. It is possible to show the presence of a magnetic field around a current-carrying conductor. A compass may be used to show that magnetic flux lines around a conductor are circular in shape. The direction of the current flow and the magnetic flux lines can be shown by using the *left-hand rule* of magnetic flux. A conductor is held in the left hand as shown in Figure 1-2. The thumb points in the direction of electron current flow from negative to positive. The fingers then encircle the conductor in the direction of the magnetic flux lines.

The circular magnetic field is stronger near the conductor and becomes weaker at a greater distance. A cross-sectional end view of a conductor with current flowing toward the observer is shown in Figure 1-3. Current flow toward the observer is shown by a circle with a dot in the center. Notice that the direction of the magnetic flux lines is clockwise, as verified by using the left-hand rule.

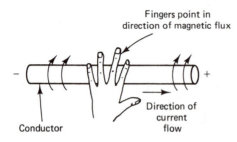

Fingers point in
direction of magnetic flux

Direction of
current
flow

Conductor

Figure 1-2   Left-hand rule of magnetic flux.

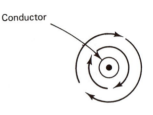

Conductor

Figure 1-3   Cross section of conductor with current flow toward the observer.

When the direction of current flow through a conductor is reversed, the direction of the magnetic lines of force is also reversed. The cross-sectional end view of a conductor in Figure 1-4 shows current flow in a direction away from the observer. Notice that the direction of the magnetic lines of force is now counterclockwise.

The presence of magnetic lines of force around a current-carrying conductor can be observed by using a compass. When a compass is moved around the outside of a conductor, its needle will align itself *tangent* to the lines of force as shown in Figure 1-5.

Figure 1-4   Cross section of conductor with current flow away from the observer.

Figure 1-5   Compass aligns tangent to magnetic force lines.

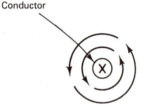

Conductor

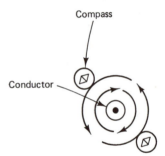

Compass

Conductor

When current flow is in the opposite direction, the compass polarity reverses but remains tangent to the conductor.

Magnetic field around a coil. The magnetic field around one loop of wire is shown in Figure 1-6. Magnetic flux lines extend around the conductor as shown when current passes through the loop. Inside the loop, the magnetic flux is in one direction. When many loops are joined together to form a coil, the magnetic flux lines surround the coil as shown in Figure 1-7. The field around a coil is much stronger than the field of one loop of wire. The field around a coil is similar in shape to the field around a bar magnet. A coil that has an iron or steel core inside it is called an *electromagnet*. The purpose of a core is to increase the magnetic flux density of a coil, thereby increasing its strength.

Figure 1-7 Magnetic field around a coil: (a) coil of wire showing current flow; (b) lines of force combine around two loops that are parallel; (c) cross section of coil showing lines of force.

Figure 1-6 Magnetic field around a loop of wire.

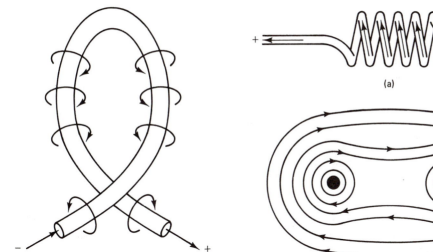

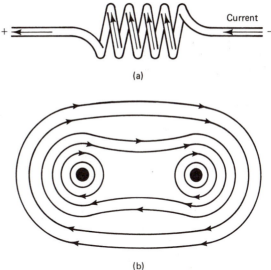

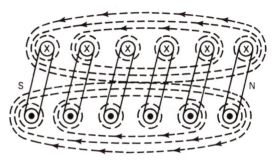

Electromagnets.   Electromagnets are produced when current flows through a coil of wire as shown in Figure 1-8. Almost all electrical machines have electromagnetic coils. The north pole of a coil of wire is the end where the lines of force exit, while the south polarity is the end where the lines of force enter the coil. To find the north pole of a coil, use the *left-hand rule for polarity,* as shown in Figure 1-9. Grasp the coil with the left hand. Point the fingers in the direction of current flow through the coil, and the thumb will point to the north polarity of the coil.

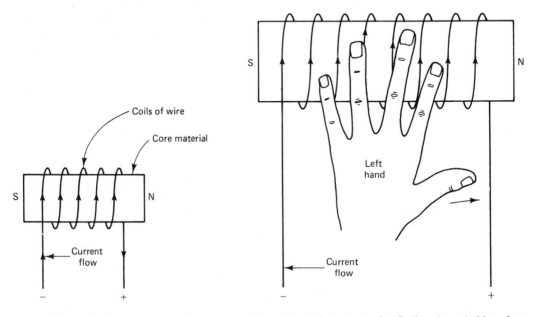

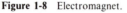

Figure 1-8 Electromagnet.

Figure 1-9 Left-hand rule for finding the polarities of an electromagnet.

When the polarity of the voltage source is reversed, the magnetic poles of the coil reverse. The poles of an electromagnet can be checked by placing a compass near a pole of the electromagnet. The north-seeking pole of the compass will point toward the north pole of the coil.

Electromagnets ordinarily have several turns of wire wound around a soft-iron core. An electrical power source is connected to the ends of the turns of wire. Thus, when current flows through the wire, magnetic polarities are produced at the ends of the soft-iron core. The three basic parts of an electromagnet are (1) an iron core, (2) wire windings, and (3) an electrical power source. Electromagnetism is made possible by electrical current flow which produces a magnetic field.

Magnetic strength of electromagnets.   The magnetic strength of an electromagnet depends on three factors: (1) the amount of current passing through the coil, (2) the number of turns of wire, and (3) the type of core material. The number of magnetic lines of force is increased by increasing the current, by increasing the number

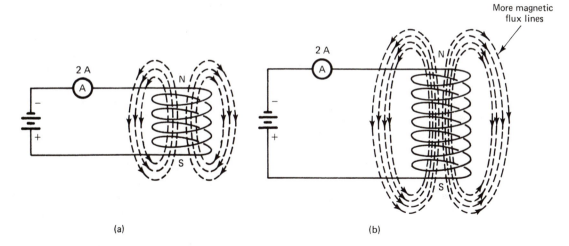

Figure 1-10 Effect of ampere-turns on magnetic field strength: (a) five turns, 2 amperes = 10 ampere-turns; (b) eight turns, 2 amperes = 16 ampere-turns.

of turns of wire, or by using a more desirable type of core material. The magnetic strength of electromagnets is determined by the *ampere-turns* of each coil. The number of ampere-turns is equal to the current in amperes multiplied by the number of turns of wire ($I \times N$). For example, 200 ampere-turns is produced by 2 amperes of current through a 100-turn coil. One ampere of current through a 200-turn coil would produce the same magnetic field strength. Figure 1-10 shows how the magnetic field strength of an electromagnet changes with the number of ampere-turns.

The magnetic field strength of an electromagnet also depends on the type of core material. Cores are usually made of soft iron or steel materials which transfer a magnetic field better than air or other nonmagnetic materials. Iron cores increase the *flux density* of an electromagnet, as shown in Figure 1-11.

Figure 1-11 Effect of an iron core on magnetic strength: (a) coil without core; (b) coil with core.

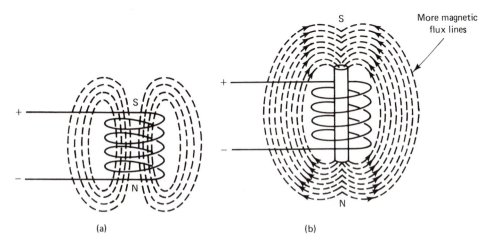

7

An electromagnet loses its field strength when current ceases to flow. However, an electromagnet's core retains a small amount of magnetic strength after current stops flowing. This is called *residual magnetism* or "leftover" magnetism. Residual magnetism can be reduced by using soft-iron cores or increased by using hard-steel core materials. Residual magnetism is extremely important in the operation of some types of electrical generators.

In many ways, electromagnetism is similar to magnetism produced by natural magnets. However, the main advantage of electromagnetism is that it is easily controlled. It is easy to increase the strength of an electromagnet by increasing the current flow through the coil. The second way to increase the strength of an electromagnet is to have more turns of wire around the core. A greater number of turns produces more magnetic lines of force around the electromagnet. The strength of an electromagnet is also affected by the type of core material used. Different alloys of iron are used to make the cores of electromagnets, some of which aid in the development of magnetic lines of force to a greater extent. Other types of core materials offer greater opposition to the development of magnetic flux around an electromagnet.

Ohm's law for magnetic circuits. Ohm's law for electrical circuits is familiar to electrical technicians. A similar relationship exists in magnetic circuits. Magnetic circuits have *magnetomotive force* (MMF), *magnetic flux* (Φ), and *reluctance* ($\Re$). MMF is the force that causes a magnetic flux to be developed. Magnetic flux refers to the lines of force around a magnetic material. Reluctance is the opposition to the development of a magnetic flux. These terms may be compared to voltage, current, and resistance in electrical circuits, as shown in Figure 1-12. When MMF increases, magnetic flux increases. Remember that in an electrical circuit, when voltage increases, current increases. When resistance in an electrical circuit increases, current decreases. When reluctance of a magnetic circuit increases, magnetic flux decreases. The relationship of these magnetic terms in electrical machinery is very important in terms of their operational characteristics.

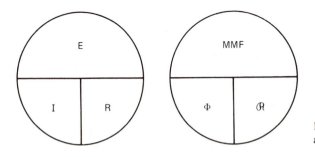

Figure 1-12 Relationship of magnetic and electrical terms.

The domain theory of magnetism. A theory of magnetism was presented in the nineteenth century by a German scientist named Wilhelm Weber. Weber's theory of magnetism was called the molecular theory and dealt with the alignment of molecules in magnetic materials. Weber felt that molecules were aligned in an *orderly* arrangement in magnetic materials. In nonmagnetic materials, he thought that molecules were arranged in a *random* pattern.

Weber's theory has now been modified somewhat to become the domain theory of magnetism. This theory deals with the alignment of the "domains" of materials rather than molecules. A domain is defined as a group of atoms (about 10^{15} atoms) that have electromagnetic characteristics. The rotation of electrons around the nucleus of these atoms is important. As negatively charged electrons orbit around the nucleus of atoms, their electrical charge also moves. This moving electrical field causes a magnetic field to be produced. The polarity of the magnetic field is determined by the direction of electron rotation.

The domains of magnetic materials are atoms which are grouped together whose electrons are believed to spin in the same direction. This produces a magnetic field due to electrical charge movement. Figure 1-13 shows the arrangement of domains in magnetic, nonmagnetic, and partially magnetized materials. In nonmagnetic materials, some of the electrons spin in one direction and some in the opposite direction, so their charges cancel each other. No magnetic field is produced since the charges cancel each other. Electron rotation of the domains in magnetic materials is in the same direction. This causes the domains to act like tiny magnets that align to produce a composite magnetic field.

(a)

(b)

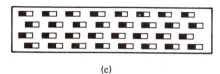

(c)

Figure 1-13 Domain theory of magnetism: (a) unmagnetized; (b) slightly magnetized; (c) fully magnetized—saturation.

Electricity produced by magnetism. A scientist named Michael Faraday discovered in the early 1830s that electricity can be produced from magnetism. He found that if a magnet is placed inside a coil of wire, electrical current is produced when the magnet is *moved*. Faraday found that electrical current is caused by relative motion between a conductor and magnetic field.

Faraday's law is stated simply as follows: When a conductor moves across the lines of force of a magnetic field, electrons will flow through the conductor in one direction. When the conductor moves across the magnetic lines of force in the opposite direction, electrons will flow through the conductor in the opposite direction. This law is the principle of electrical power generation and basic operation of mechanical generators.

Current flow through a conductor which is placed inside a magnetic field occurs only when there is relative motion between the conductor and the magnetic field. If a conductor is stopped while moving across the magnetic lines of force, current will stop flowing. The operation of electrical generators depends on conductors that pass through an electromagnetic field. This principle is called electromagnetic induction.

Electromagnetic induction.   The principle of electromagnetic induction was one of the most important discoveries in the development of modern electrical technology. Without electrical power, our lives would certainly be different. Electromagnetic induction, as the name implies, involves electricity and magnetism. When electrical conductors, such as alternator windings, are moved within a magnetic field, an electrical current is developed in the conductors. The electrical current produced in this way is called an *induced current*. A simplified illustration showing how induced electrical current is developed is shown in Figure 1-14. This principle was developed by Michael Faraday in the early nineteenth century.

A conductor is placed within the magnetic field of a horseshoe magnet so that the left side of the magnet has a north polarity (N) and the right side has a south polarity (S).

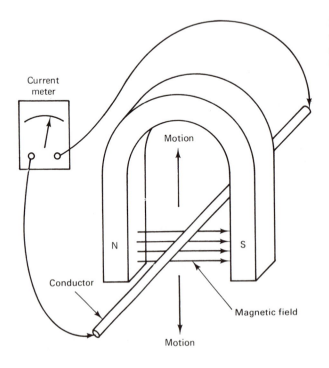

Figure 1-14   Faraday's law: Electrical current is produced when there is relative motion between a conductor and a magnetic field.

Magnetic lines of force travel from the north polarity of the magnet to the south polarity. The ends of the conductor are connected to a current meter to measure the induced current. The meter is the zero-centered type, so its needle can move either to the left or to the right. When the conductor is moved, current will flow through the conductor. Electromagnetic induction takes place whenever there is relative motion between the conductor and the magnetic field. Either the conductor can be moved through the magnetic field or the conductor can be held stationary and the magnetic field can be moved past it. Thus current will be induced whenever there is a relative motion between the conductor and magnetic field.

If the conductor shown in Figure 1-14 is moved upward, the needle of the meter will move in one direction. However, if the conductor is move downward, the needle of the meter will deflect in the other direction. This shows that the direction of movement of the conductor within the magnetic field determines the direction of current flow. In one case, the current flows through the conductor from the front of the illustration to the back. In the other situation, the current travels from the back to the front. The direction of current flow is indicated by the direction of the meter deflection. The principle demonstrated here is the basis for large-scale electrical power generation.

In order for an induced current to be developed, the conductor must have a complete path or closed circuit. The meter in Figure 1-14 was connected to the conductor to make a complete current path. If there is no closed circuit, electromagnetic induction cannot take place. It is important to remember that an induced current causes an induced electromotive force (voltage) across the ends of the conductor.

Magnetic effects.   There are several magnetic effects which are evident during the operation of electrical machines. These effects include residual magnetism, permeability, retentivity, and magnetic saturation.

Residual magnetism is an effect that is important in the operation of some types of electrical generators. Residual magnetism is the ability of electromagnetic coils to retain a small magnetic field after electrical current ceases to flow. A small magnetic field remains around an electromagnetic coil after it is deenergized. The effect of residual magnetism for self-excited dc generators will be discussed in Chapter 5.

Permeability is the ability of a magnetic material to transfer magnetic flux or the ability of a material to magnetize and demagnetize. Soft-iron material has a high permeability, since it transfers magnetic flux very easily. Soft-iron magnetizes and demagnetizes rapidly, making soft iron a good material to use in the construction of generators, motors, transformers, and other electrical machines.

A related term is *relative permeability*. This is a comparison of the permeability of a material to the permeability of air (1.0). Suppose that a material has a relative permeability of 1000. This means that the material will transfer 1000 times more magnetic flux than an equal amount of air. The highest relative permeability of materials is over 5000.

Another magnetic effect is called *retentivity*. The retentivity of a material is its ability to retain a magnetic flux after a magnetizing force is removed. Some materials will retain a magnetic flux for a long period of time, while others lose their magnetic flux almost immediately after the magnetizing force is removed.

Magnetic saturation is important in the operation of electrical machines, especially generators. Saturation is best explained by the magnetization (*B–H*) curve shown in Figure 1-15. This curve shows the relationship between a magnetizing force (*H*) and flux density (*B*). Notice that as a magnetizing force increases, so does the flux density. *Flux density (B)* is the amount of lines of force per unit area of a material. An increase in flux density occurs until *magnetic saturation* is reached. This saturation point depends on the type of core material used. At the saturation point, the maximum alignment of domains takes place in the core material.

Magnetizing force (H) is measured in *oersteds*. The base unit is the number of ampere-turns per meter of length. *Flux* density (*B*) is the amount of magnetic flux per unit area and is measured in *gauss* per square centimeter of area.

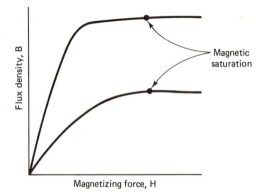

Figure 1-15 Magnetization curves of two different materials.

ELECTRICAL MEASUREMENT

Most nations today use the metric system of measurement. In the United States, the National Bureau of Standards began a study in 1968 to determine the feasibility and costs of converting industry and everyday activity to the metric system. Today, this conversion is still taking place. English units of measurement are used for many electrical machinery applications in the United States today.

The units of the metric system are decimal measures based on the kilogram and the meter. Although this system is very simple, several countries have been slow to adopt it. The United States has been one of these reluctant countries due to the complexity of a complete changeover of measurement systems.

Most measurement is based on the International System of Units (SI). The basic units of this system are the meter, kilogram, second, and ampere (MKSA). These are the units of length, mass, time, and electrical current. Other systems, such as the meter–kilogram–second (MKS) and centimeter–gram–second (CGS), recognize only three base units. However, these two systems are closely associated with the metric system.

There are several derived units that are used extensively for electrical and other related measurements. The electrical units that are now used are part of the International System of Units (SI) based on the meter–kilogram–second–ampere (MKSA) system. The International System of Derived Units is shown in Figure 1-16.

The coordination necessary to develop a standard system of electrical units is very complex. An International Advisory Committee on Electricity makes recommendations to the International Committee on Weights and Measures. Final authority is held by the General Conference on Weights and Measures. The laboratory associated with the International System is the International Bureau of Weights and Measures, located near Paris, France. Several laboratories in different countries cooperate in the process of standardizing units of measurement. One such laboratory is the National Bureau of Standards in the United States.

Figure 1-16 International system of derived units.

Measurement Quantity	SI Unit
Area	square meter
Volume	cubic meter
Frequency	hertz
Density	kilogram per cubic meter
Velocity	meter per second
Acceleration	meter per second per second
Force	newton
Pressure	pascal (newton per square meter)
Work (energy), quantity of heat	joule
Power (mechanical, electrical)	watt
Electrical charge	coulomb
Permeability	henry per meter
Permittivity	farad per meter
Voltage, potential difference, electromotive force	volt
Electric flux density, displacement	coulomb per square meter
Electric field strength	volt per meter
Resistance	ohm
Capacitance	farad
Inductance	henry
Magnetic flux	weber
Magnetic flux density (magnetic induction)	tesla
Magnetic field strength (magnetic intensity)	ampere per meter
Magnetomotive force	ampere
Magnetic permeability	henry per meter
Luminous flux	lumen
Luminance	candela per square meter
Illumination	lux

CONVERSION OF ELECTRICAL UNITS

Sometimes it is necessary to make conversions of electrical units so that very large or very small numbers may be avoided. For this reason, decimal multiples and submultiples of the basic units have been developed by using standard prefixes. These standard prefixes are shown in Figure 1-17. As an example, we may express 1000 volts as a 1 kilovolt or 0.001 ampere as 1 milliampere.

A chart is shown in Figure 1-18 that can be used for converting from one unit to another. To use this conversion chart, follow these simple steps:

1. Find the position of the unit as expressed in its original form.

Prefix	Symbol	Factor by Which the Unit is Multiplied
exa	E	$1,000,000,000,000,000,000 = 10^{18}$
peta	P	$1,000,000,000,000,000 = 10^{15}$
tera	T	$1,000,000,000,000 = 10^{12}$
giga	G	$1,000,000,000 = 10^{9}$
mega	M	$1,000,000 = 10^{6}$
kilo	k	$1,000 = 10^{3}$
hecto	h	$100 = 10^{2}$
deka	da	$10 = 10^{1}$
deci	d	$0.1 = 10^{-1}$
centi	c	$0.01 = 10^{-2}$
milli	m	$0.001 = 10^{-3}$
micro	μ	$0.000001 = 10^{-6}$
nano	n	$0.000000001 = 10^{-9}$
pico	p	$0.000000000001 = 10^{-12}$
femto	f	$0.000000000000001 = 10^{-15}$
atto	a	$0.000000000000000001 = 10^{-18}$

Figure 1-17 Standard prefixes.

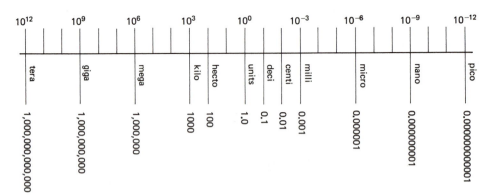

Figure 1-18 Conversion chart for small and large numbers.

2. Find the position of the unit to which you are converting.
3. Write the original number as a whole number or in scientific notation.
4. Shift the decimal point the appropriate number of units in the direction of the term to which you are converting, and count the difference in decimal multiples from one unit to the other.

The use of this step-by-step procedure is illustrated in the following examples:

A. Convert 100 microamperes to amperes.
  1. 100 microamperes (μA)
  2. _____ amperes (A)
  3. 100 μA or 100×10^{-6}
  4. 0.0001 A (decimal shifted six units to the left)
B. Convert 20,000 ohms to kilohms.
  1. 20,000 ohms (Ω)
  2. _____ kilohms (kΩ)
  3. 20,000 Ω or $20,000 \times 10^{0}$
  4. 20 kΩ (decimal shifted three units to the left)
C. Convert 10 milliamperes to microamperes.
  1. 10 milliamperes (mA)
  2. _____ microamperes (μA)
  3. 10 mA or 10×10^{-3}
  4. 10,000 μA (decimal shifted three units to the right)

Another measurement conversion method is the conversion of U.S. system units to SI units, and vice versa. These conversions are shown in Appendix 1.

ELECTRICAL POWER SYSTEMS

The organization of this text follows a simple model of an electrical power "system." Rotating electrical machines and transformers are used primarily for power-related applications. Thus it is quite logical to study electrical machines and transformers using the framework of an "electrical power system." The electrical power system model used for this book is shown in Figure 1-19.

The word "system" is commonly defined as "an organization of parts that are connected together to form a complete unit." There are a wide variety of different systems used today. An electrical power system, for example, is needed to produce electrical energy and distribute it to each consumer. Also, mechanical systems are needed to hold objects for machining operations and to move them physically in production lines in industry.

Each system obviously has a number of unique features or characteristics that distinguish it from other systems. More important, however, there is a common set of

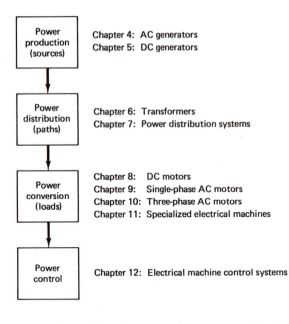

Power production (sources)	Chapter 4: AC generators Chapter 5: DC generators
Power distribution (paths)	Chapter 6: Transformers Chapter 7: Power distribution systems
Power conversion (loads)	Chapter 8: DC motors Chapter 9: Single-phase AC motors Chapter 10: Three-phase AC motors Chapter 11: Specialized electrical machines
Power control	Chapter 12: Electrical machine control systems

Figure 1-19 Electrical power system model.

parts found in each system. These parts play the same basic role in all systems. The terms "energy source," "distribution path," "control," "load," and "indicator" are used to describe the various system parts. A block diagram of these basic parts of the system is shown in Figure 1-20.

Each block of a basic system has a specific role to play in the operation of the system. Hundreds of components are sometimes needed to achieve a specific block function. Regardless of the complexity of the system, each block must achieve its function in order for the system to be operational.

The *energy source* of a system is responsible for converting energy of one form into another form. Heat, light, sound, chemical, nuclear, and mechanical energy are typical sources of energy.

The *distribution path* of a system is somewhat simplified compared with other system functions. This part of the system provides a path for the transfer of energy. It starts with the energy source and continues through the system to the load. In some

Figure 1-20 Block diagram of a basic electrical system.

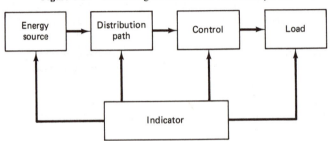

cases, the path may be a single feed line or electrical conductor connected between the source and the load. In other systems, there may be a supply line between the source and the load, and a return line from the load to the source. There may also be a number of alternative paths within a complete system. These paths may be series connected to a number of load devices, or parallel connected to many independent devices.

The *control* section of a system is by far the most complex part of the entire system. In its simplest form, control is achieved when a system is turned on or off. Control of this type can take place anywhere between the source and the load device. The term *full control* is used to describe this operation. In addition to this type of control, a system may also employ some type of *partial control*. Partial control usually causes some type of an operational change in the system other than an on or off condition. Changes in electric current, hydraulic pressure, and airflow are some of the system alterations achieved by partial control.

The *load* of a system refers to a specific part or number of parts designed to produce some form of work. Work occurs when energy goes through a conversion from one form to another. Heat, light, chemical action, sound, and mechanical motion are some of the forms of work produced by a load device. As a general rule, a large portion of the energy produced by the source is consumed by the load during its operation. The load is typically a prevalent part of the entire system because of its work function.

The *indicator* of a system is designed primarily to display certain operating conditions at various points throughout the system. In some systems the indicator is an optional part, while in others it is an essential part in the operation of the system. An indicator may only temporarily be attached to the system to make measurements. Test lights, meters, oscilloscopes, chart recorders, digital display instruments, and pressure gauges are some of the common indicators used.

An electrical power system supplies energy to our homes and industrial buildings. Figure 1-21 shows a sketch of a simple electrical power system. The source of energy may be derived from coal, oil, natural gas, atomic fuel, or moving water. This type of energy is needed to produce mechanical energy, which in turn develops the rotary motion of a turbine. Massive alternators are then rotated by the turbine to produce alternating-current electrical energy.

In an electrical power system, the distribution path is achieved by a large number of electrical conductors. Copper wire and aluminum wire are used more frequently today than any other type of conductor. The distribution path of an electrical power system often becomes very complex.

The control function of an electrical power system is achieved by a variety of different devices. Full control, for example, is accomplished by three distinct types of circuit-interrupting equipment. These include switches, circuit breakers, and fuses. Each piece of equipment must be designed to pass and interrupt specific values of current. Making and breaking distribution path connections with a minimum of arc is one of the most important characteristics of this type of equipment.

Partial control of an electrical power system is achieved primarily by circuit resistance, inductance, and capacitance. The resistance of a long-distance transmission

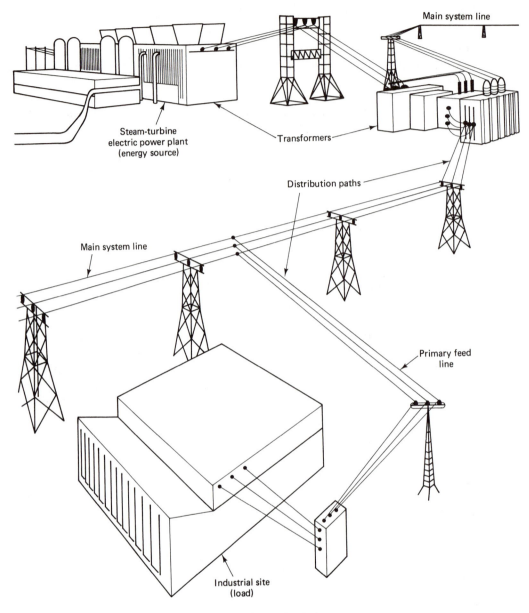

Figure 1-21 Simple electrical power system.

line is not a desirable feature. It must be minimized to avoid unnecessary system losses. Any power loss that occurs in the transmission line is wasted energy that cannot be utilized by the load of the system.

To minimize losses in an electrical power system, transformers are used at strategic locations throughout the system. These partial-control devices are designed to initially set up the source voltage to a high value. Through this process, the source current is proportionally reduced in value. Since the power loss of a transmission line is based on the square of the current times the line resistance ($I^2 \times R$), losses can be reduced to a reasonable value through this process.

Transformers are also used to lower system voltages to operational values near the load. This action of a transformer is described as its step-down function. When the source voltage is reduced to a lower value, the source current is proportionally increased in value. Through the use of transformers, transmission-line losses can be held to a minimum, thus causing increased system efficiency.

Capacitors are used in electrical power systems primarily as power-factor correcting devices. In this function, a capacitor can be used to cause the current to lead the voltage by a certain amount. This effect is used to reduce the lagging current condition caused by transformer inductance and large motor loads. When the alternating current and voltage of an electrical power system are kept in phase, the operating efficiency of the system is improved.

The load of an electrical power system is usually quite complex. As a composite, it includes everything that utilizes electrical energy from the source. As a general rule, the load is divided into four distinct parts. This includes residential, commercial, industrial, and other uses, such as street lighting. The composite load of an electrical power system is subjected to change hourly, daily, and seasonally.

The average person is probably more familiar with the load part of an electrical power system than with any of its other parts. This represents the part of the system that actually does work for us. Motors, lamps, electric ovens, welders, and power tools are some of the common load devices used. Loads are frequently classified according to the type of work they produce. This includes light, heat, sound, electromechanical changes, and chemical action. Loads are also classified according to the predominant electrical characteristics involved. This includes resistive, capacitive, and inductive loads.

The indicator of an electrical power system is designed to show the presence of electrical energy at various places or to measure different electrical quantities. Meters, oscilloscopes, chart recording instruments, and digital display devices are some of the indicators used in this type of system today. Voltage indicators, line-current indicators, power-factor indicators, temperature gauges, water-level meters, pressure gauges, and chart recorders are found in electrical power systems. Indicators of this type are designed to provide an abundance of system operating information.

REVIEW

1.1. Why is the study of electrical machines and power systems becoming more important?

1.2. How have the applications of electrical machines expanded in recent years?

1.3. How did widespread alternating-current (ac) power distribution systems come into existence?

1.4. List several applications of electric motors.

1.5. Why are magnetism and electromagnetism important topics in the study of electrical machines and power systems?

1.6. What are the "laws of magnetism?"

1.7. What is a magnetic field?

1.8. Draw a cross-sectional view of two conductors to show how a magnetic field is developed around current-carrying conductors.

1.9. How is an electromagnet produced?

1.10. How is the magnetic strength of an electromagnet affected?

1.11. What is residual magnetism?

1.12. Define the following magnetic terms: **(a)** magnetomotive force, **(b)** magnetic flux, and **(c)** reluctance.

1.13. Briefly explain the domain theory of magnetism.

1.14. What is Faraday's law, and why is it important in the study of electrical machines?

1.15. What is electromagnetic induction?

1.16. Discuss the following magnetic terms: **(a)** permeability, **(b)** retentivity, **(c)** magnetic saturation, and **(d)** B–H curve.

1.17. Discuss the use of English and SI units of measurement.

1.18. Discuss the electrical power system model used in the organization of this book.

1.19. What are the five parts of a basic electrical system?

PROBLEMS

1.1 Perform each of the following unit conversions.

(a) 5.31 megohms to ohms

(b) 6320 watts to kilowatts

(c) 2180 milliamperes to amperes

(d) 926 microvolts to volts

(e) 8520 volts to kilovolts

(f) 50 microamperes to milliamperes

(g) 2,526,000 ohms to megohms

(h) 8200 watts to kilowatts

(i) 6.38 volts to millivolts

(j) 3160 microamperes to amperes

(k) 15,500 ohms to kilohms

(l) 0.053 ampere to milliamperes

1.2. Perform the following conversions using the tables in Appendix 1.

(a) 3 cubic feet to cubic meters

(b) 8 cubic centimeters to cubic inches

(c) 30 miles to kilometers

(d) 12 cups to liters

(e) 6 gallons to liters

(f) 8 liters to quarts

(g) 3 kilograms to pounds

(h) 40 ounces to grams

(i) 5 grams to ounces

(j) 3 kilometers to miles

(k) 2 centimeters to inches

(l) 6 meters to inches

(m) 3 liters to milliliters

(n) 5200 grams to kilograms

(o) 2 centimeters to millimeters

(p) 3 meters to centimeters

(q) 62° F to °C

(r) 39°C to °F

TWO

Construction and Basic Characteristics of Electrical Machines

Rotating electrical machines accomplish electromechanical energy conversion. Generators convert mechanical energy into electrical energy, while motors convert an electrical energy input into a mechanical energy output. Generators and motors have basic construction characteristics which are common among many types of machines. The functions of various machines differ even though their construction is similar. Generators have rotary motion supplied by prime movers which provide mechanical energy input. Relative motion between the conductors and a magnetic field of generators produces an electrical energy output. Motors have electrical energy supplied to their windings and a magnetic field that develops an electromagnetic interaction to produce mechanical energy or torque. This chapter discusses the construction and basic characteristics of electrical machines.

ROTATING MACHINERY CONSTRUCTION

The construction of most rotating electrical machines is somewhat similar. Most machines have a stationary part called the *stator* and a rotating set of conductors called the *rotor*. The stator consists of a *yoke* or *frame* which serves as a support and a metallic path for magnetic flux developed in a machine. Machine stator assemblies are shown in Figure 2-1. Figure 2-1(a) shows a cutaway view of a small gear-reduction motor. Larger motor stator assemblies are shown in Figure 2-1(b) and (c). Notice the design of the mounting brackets on the stator of each type.

(a)   (b)

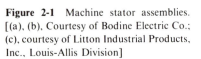

Figure 2-1 Machine stator assemblies. [(a), (b), Courtesy of Bodine Electric Co.; (c), courtesy of Litton Industrial Products, Inc., Louis-Allis Division]

(c)

Field poles and windings.   Rotating machines also have *field poles* which are part of the stator assembly. Field poles are constructed of laminated sheets of steel and secured to the machine frame. They are usually curved on the portion near the rotor to provide a low-reluctance path for magnetic flux. The *field windings* or *field coils* are placed around the poles as shown in cutaway view of Figure 2-2. The field coils are electromagnets that develop an electromagnetic field interaction with the rotor to generate a voltage or to produce torque in a machine.

There are two major types of electromagnetic field coils used in the construction of electrical machines. These are the (1) *concentrated* or salient coils and (2) *distributive* field coils. As a general rule, salient coils are used more frequently in dc motors and generators, while distributive coils are found in most ac applications.

Figure 2-2 Field coils placed around laminated field poles (cutaway view). (Courtesy of Bodine Electric Co.)

Figure 2-3 shows examples of these two types of coil windings. Coil windings are insulated from each other and from the field poles. Figure 2-4 shows distributive windings of stator assemblies. Figure 2-4(a) shows a large machine stator assembly and Figure 2-4(b) shows the stator of a smaller motor being soldered. Figure 2-4(c) shows a heavily insulated distributive wound stator of a small servomotor. A stator assembly of a large machine with distributive windings is illustrated in Figure 2-4(d).

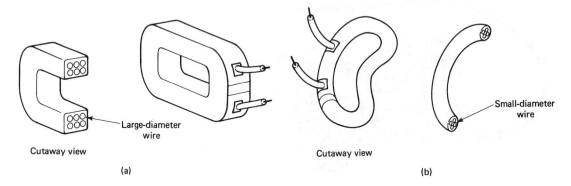

Figure 2-3   Coil windings: (a) concentrated or salient type; (b) distributive type.

Figure 2-4   Distributive windings used on machine stator assemblies. [(a), (b), Courtesy of Litton Industrial Products, Inc., Louis-Allis Division.; (c), courtesy of Bodine Electric Co.; (d), courtesy of Lima Electric Co., Inc.]

Rotor construction.   In the study of electrical machines, there is a need to understand the electromagnetic fields produced by the rotating section of a motor or a generator. This section is called the armature or rotor and its coils are also of either the salient or distributive type. Some types of machines use solid metal rotors called squirrel-cage rotors. Types of rotor assemblies are shown in Figure 2-5.

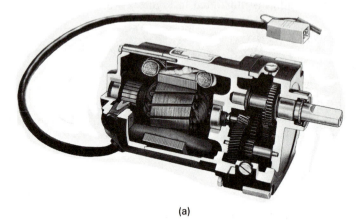

(a)

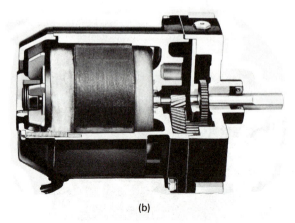

(b)

Figure 2-5 Machine rotor assemblies: (a) with windings and split-ring commutator and brushes; (b) with laminated solid rotor. [(a), Courtesy of Litton Industrial Products, Inc., Louis-Allis Division.; (b), courtesy of Bodine Electric Co.]

Slip rings, split rings, and brushes.   In order for electrical energy to be supplied to a rotating device such as the armature, some sort of sliding brush contact must be established. Sliding brush contacts are either *slip rings* or *split rings*. Slip rings are constructed of a cylinder of insulating material with two separate solid metal rings glued to it. Sliding *brushes* made of carbon and graphite ride on the metal rings and permit application or extraction of electrical energy from the rings during rotation. The split-ring commutator is similar to the slip ring except a solid metal ring is cut into two or more separate sections. As a general rule, slip rings are used in ac motors and generators, while dc machines employ the split-ring commutating device. The gap or split in the commutator is kept at a minimum to reduce sparking of the brushes. Slip rings and split rings are shown in Figure 2-6.

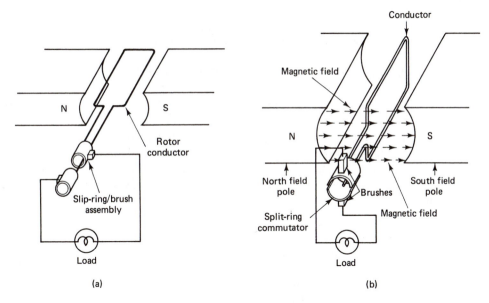

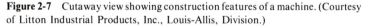

Figure 2-6 (a) Slip rings; (b) split rings.

Other machine parts. There are several other parts used in the construction of rotating machines. Among these are the rotor *shaft,* which rotates between a set of *bearings.* Bearings may be either the ball, roller, or sleeve type. Bearing *seals,* often made of felt material, are used to keep lubricant around the bearing and keep dirt out. A rotor *core* is usually constructed of laminated steel to provide a low-reluctance magnetic path between the field poles of a machine and to reduce eddy currents. Internal and external electrical *connections* provide a means of delivering or extracting electrical energy. Several of the construction features of electrical machines are shown in Figure 2-7.

Figure 2-7 Cutaway view showing construction features of a machine. (Courtesy of Litton Industrial Products, Inc., Louis-Allis, Division.)

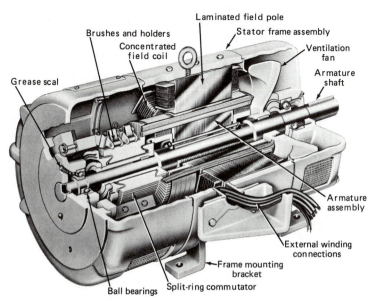

STATOR WINDING DESIGN OF ELECTRICAL MACHINES

The electromagnetic coils used to produce a magnetic field in rotating electrical machines are called *field coils*. In a *two-pole* machine two coils are used to develop a magnetic circuit as shown in Figure 2-8. The coils are wound around laminated metal cores called *pole pieces*. The strength of the magnetic field developed around the pole pieces is affected by their magnetic properties and the properties of the stator assembly. The primary purposes of the field coils is to develop a magnetic circuit for an electrical machine.

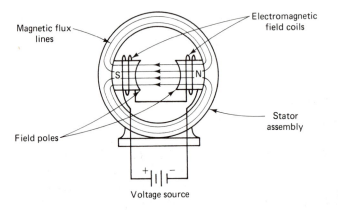

Figure 2-8 Two-pole electrical machine.

Winding polarities.   Basic magnetic laws apply to the development of magnetic circuits in electrical machines. The coil shown in Figure 2-9 is energized by a direct-current (dc) voltage source. If a compass is placed along a central axis near the energized coil, a polarity is developed on either end. If the power source polarity is reversed, the magnetic polarities are also reversed.

The same principle applies to electromagnetic coils used in electrical machines. Figure 2-10 shows an electromagnetic coil energized by a dc power source. The dark band on the lower right section indicates the beginning of the coil winding for reference purposes. The positive side of the power source connects to the end of the winding and

Figure 2-9 Coil energized by a dc source.

Figure 2-10 Electromagnetic coil energized by a dc source.

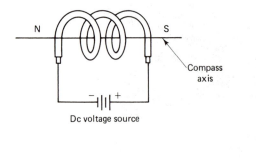

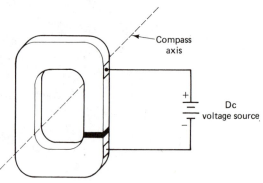

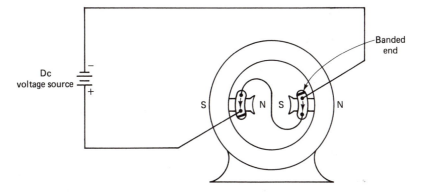

Figure 2-11   Proper method of connecting the windings of a two-pole machine.

the negative side to the beginning. If the power source polarity is reversed, the magnetic polarity indicated along the compass axis of the figure is also reversed.

Polarities developed in an electrical machine must be a result of proper coil connections. Figure 2-11 shows a method of connecting the windings of a two-pole electrical machine. Dark bands are used to represent the beginnings of each of the windings. Notice that these coils are connected in series to form opposite polarities. The wire that connects the two coils extends from the end of one winding to the end of the next winding. The assumption here is that current flow from beginning to end through a coil produces an internal south polarity and current flow from end to beginning cause an internal north polarity. The current directions are shown on the diagram with arrows. If a compass is used to check polarities, the external and internal polarities would be as shown. Notice that in either case, the polarities are opposite. Opposite polarities are used to develop the proper internal magnetic circuit for an electrical machine.

Multiple field windings.   Many rotating electrical machines have more than two poles. There must always be an even number of field windings, such as a four-pole or six-pole machine. Additional poles in an electrical machine provide a better magnetic circuit by reducing the reluctance of the magnetic flux lines. Reduced reluctance allows a more efficient internal transfer of magnetic flux through a machine. The practical effect of adding more field windings is to increase generator output capability and motor torque characteristics.

A pictorial diagram of a four-pole electrical machine is shown in Figure 2-12(a). The coils are connected in series to form the proper alternate north and south polarities. Notice the direction of current flow through each coil. The internal polarities produce an alternate north-south arrangement due to the direction of current flow through each coil. The schematic shown in Figure 2-12(b) may be helpful in understanding the proper field winding connections for a four-pole electrical machine with a series circuit design. Figure 2-13(a) shows a parallel connection for a four-pole machine with a schematic representation in Figure 2-13(b). Parallel

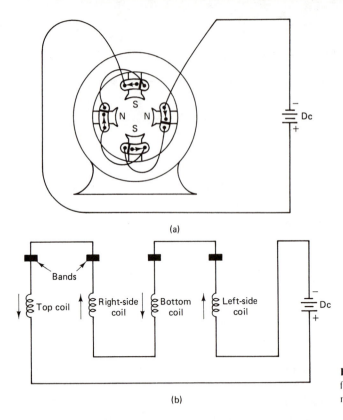

(a)

Bands

Top coil | Right-side coil | Bottom coil | Left-side coil

Dc

(b)

Figure 2-12 Series coil connections for a four-pole machine: (a) pictorial; (b) schematic.

Figure 2-13 Parallel coil connections for a four-pole machine: (a) pictorial diagram; (b) schematic diagram.

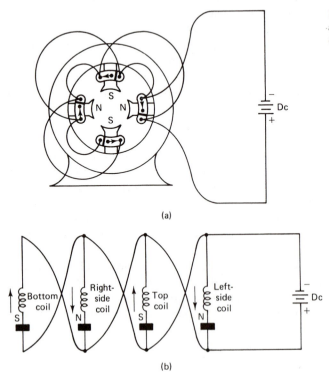

(a)

Bottom coil | Right-side coil | Top coil | Left-side coil

S | N | S | N

Dc

(b)

Figure 2-14 Eight-pole stator assembly. (Courtesy of Bodine Electric Co.)

connections supply full source voltage across each coil, while series connections distribute the source voltage in four equal parts across the coils. An eight-pole concentrated winding stator assembly is shown in Figure 2-14.

ROTOR WINDING DESIGN OF ELECTRICAL MACHINES

Several rotating electrical machines use wound-rotor assemblies. These assemblies are ordinarily called *armatures*. Armature coils are wound into slots on laminated metal cores as shown in Figure 2-15. Figure 2-16 shows the wound-rotor assembly of a four-pole rotor with concentrated windings. The coils make contact with either a split-ring commutator or a slip-ring/brush assembly. The metallic core material provides a low-reluctance magnetic circuit for the armature.

The output characteristics of an electrical machine depend on the proper design of the armature circuit. There are two common methods used to connect armature windings to split-ring commutators of dc machines. These methods are called (1) lap

Figure 2-16 Four-pole wound-rotor assembly. (Courtesy of Lima Electric Co., Inc.)

Figure 2-15 Armature coils wound onto a laminated metal core.

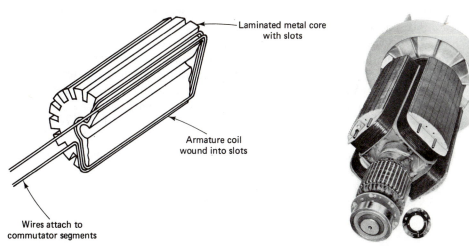

Laminated metal core with slots

Armature coil wound into slots

Wires attach to commutator segments

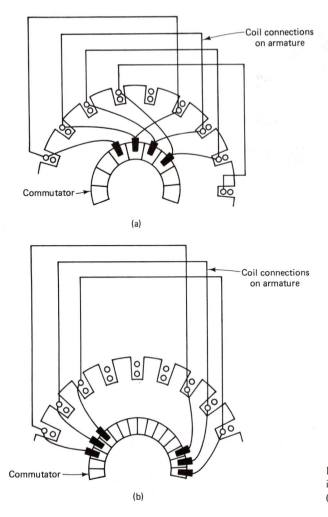

(a)

(b)

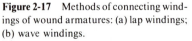

Figure 2-17   Methods of connecting windings of wound armatures: (a) lap windings; (b) wave windings.

windings and (2) wave windings. Pictorial representations of these winding methods are shown in Figure 2-17.

Lap windings [see Figure 2-17(a)] are wound so that one coil overlaps the previous coil. The beginning and end of a lap-wound coil connect to adjacent split-ring commutator segments. The commutator segments make connection to adjacent coils. The result is to place the coils in a *parallel* connection. In lap-wound armatures, there are the same number of parallel paths through the winding circuit as there are field poles. For each two parallel paths, there is one set of brushes. Lap windings are used where current capacity, rather than voltage, is the main consideration.

Wave windings [see Figure 2-17(b)] have the beginning of one coil connected to a split-ring commutator segment a distance equal to *two poles* away from the segment where the end of the coil is connected. The result is to place the coils in a series connection. In wave-wound armatures, there are only *two* parallel paths for current flow regardless of the number of field poles. Wave windings are used where high voltage, rather than current capacity, is the primary design factor.

Armature Ratings

Armature windings of electrical machines have two or more parallel paths in an even-numbered arrangement. Each path has a series-connected group of coils, in both lap-wound and wave-wound armatures. The voltage rating of the armature circuit is determined by the number of series-connected coils per path. The current rating of the armature circuit is dependent on the current-carrying capacity of the conductors in a series-connected group of coils. Increasing the number of parallel paths causes the current rating to increase. Since an armature circuit has a fixed number of coils, current rating is increased at the expense of voltage rating. This relationship is analogous to a group of batteries connected together. Connecting batteries in series increases voltage and connecting them in parallel increases current capacity (see Figure 2-18). Notice in Figure 2-18 that the power rating of each circuit arrangement is the same. Since power equals the product of voltage and current ($V \times I$), power rating is the same regardless of circuit configuration. This same relationship is true for armature winding connections. The power rating is determined by the voltage and current ratings of individual coils in a path. The only method that can be used to increase power rating of an armature winding is to design an armature that has more conductors and is larger in physical size. Thus higher power ratings require armatures that are larger in physical size.

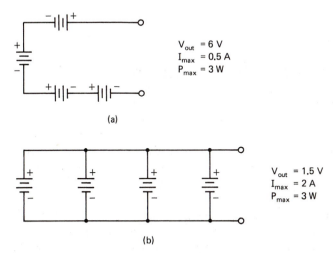

$$V_{out} = 6\ V$$
$$I_{max} = 0.5\ A$$
$$P_{max} = 3\ W$$

(a)

$$V_{out} = 1.5\ V$$
$$I_{max} = 2\ A$$
$$P_{max} = 3\ W$$

(b)

Figure 2-18 Analogy of armature windings and batteries in (a) series and (b) parallel connections. Each battery is rated at 1.5 V, 0.5 A.

Example of average voltage induced in an armature circuit. The voltage induced into the armature circuit of a dc generator is produced by many coils connected in series. To determine the voltage developed across the brushes of a dc generator, the average induced voltage per conductor in one-fourth revolution (90°) must be found. Rotation of a coil from 0 to 90° is from minimum magnetic flux linkage to maximum flux linkage. This average voltage (V_{avg}) for one-fourth revolution is found by the formula

$$V_{avg} = \frac{\phi}{t} \times 10^{-8} \text{ V}$$

where ϕ = number of magnetic flux lines per pole
       t = time required for 1/4 revolution
       10^{-8} = a constant

The average induced voltage per coil (V_{avg}/coil) for a complete revolution (four quarter-revolutions) may be found by applying the equation

$$V_{avg}/\text{coil} = 4\phi Ns \times 10^{-8} \text{ V}$$

where N = number of turns per coil
      s = relative speed in revolutions per second (r/s) between the coil and the magnetic field

For example, assume that the number of magnetic flux lines per pole is 5×10^6 lines, the number of turns per coil is two, and the speed is 100 r/min. The average voltage per coil is

$$V_{avg}/\text{coil} = 4 \times (5 \times 10^6 \text{ lines/pole}) \times (2 \text{ turns}) \times (100 \text{ r/min} \times \tfrac{1}{60} \text{ min/s}) \times 10^{-6}$$

$$= 4(5 \times 10^6)(2)(100 \times 0.16)(10^{-8})$$

$$= 64 \text{ V}$$

Armature Winding Design

Most armature windings have preformed coils which are inserted into the slots of a laminated metal armature core. Each coil has many turns of small insulated wire. Ordinarily, they are taped or dipped in lacquer to insulate them from the armature core material. Usually, a coil spans 180 electrical degrees and is either lap-wound or wave-wound. A coil that covers 180° is called a *full-pitch* coil and one that spans less than 180° is called a *fractional-pitch* coil. Armatures wound with a *fractional-pitch* are called *chorded windings*. Another term associated with fractional-pitch coils is *pitch factor* (p). A coil spanning 120 electrical degrees has a pitch factor of 120°/180° or $p =$ 0.667 (66.7%).

Most armature windings are *two-layer* windings having the sides of two coils inserted into one armature core slot. Many series-connected coils having two or more current paths are used to form the armature winding circuit. The two methods used to form the end connections of the individual coils are *lap windings* and *wave windings*, as discussed previously. Both types of windings have one coil side adjacent to a north field pole while the other is adjacent to a south pole. Lap windings have coil ends connected to adjacent commutator bars, while wave windings do not.

There are several methods of connecting either lap or wave windings to produce various voltage and current characteristics for dc machines. *Multiplex windings* have several sets of windings connected together in the armature core. In contrast, a *simplex winding* has only one set of coils to form a winding. If there are three winding sets connected together on an armature core, it is referred to as a *triplex winding*. Two sets of windings are called *duplex windings*. The number of winding sets is referred to as the *multiplicity* of the armature. Multiplicity affects the number of parallel paths in an armature winding circuit

Generated voltage calculation example.   To calculate the generated voltage of a dc generator, the multiplicity of the windings must be considered. The number of current paths for a lap winding is found as follows: $M \times P$, where M is the multiplicity (2 for duplex, 3 for triplex, etc.), and P is the number of poles. For wave windings: paths $= 2 \times M$. The generated voltage (V_g) of a dc generator is dependent on the number of current paths as shown by the following formula:

$$V_g = \frac{\phi ZSP}{60(\text{number of paths})}$$

where ϕ = magnetic flux per pole

$\quad Z$ = number of armature conductors (2 × number of turns)

$\quad S$ = speed of rotation in r/min

$\quad P$ = number of poles

For example, assume that a duplex, lap-wound armature is connected in an eight-pole machine, the magnetic flux per pole is 5×10^5 lines, the speed of rotation is 100 r/min, and there are 500 coils, each having two turns of wire. The following method is used to calculate the generated voltage (V_g).

1. Calculate the number of paths.

$$\text{paths} = M \times P$$
$$= 2 \times 8$$
$$= 16$$

2. Calculate the number of conductors.

$$Z = 500 \text{ coils} \times 2 \text{ turns/coil} \times 2 \text{ conductors/turn}$$
$$= 2000 \text{ conductors}$$

3. Calculate V_g.

$$V_g = \frac{\phi ZSP}{60(\text{number of paths})} \times 10^{-8}$$

$$= \frac{(5 \times 10^5)(2000)(100)(8)}{60(16)} \times 10^{-8}$$

$$= 83.33 \text{ V}$$

Lap-winding design.   For lap windings, the number of current paths in parallel is determined by the product of multiplicity and the number of poles ($M \times P$). The current flow through each armature coil is equal to the total armature current (I_T) divided by the number of paths. Lap windings require the same number of brushes as poles. The example shown in Figure 2-19 is an illustration of a four-pole lap-wound simplex winding. The circuit connections are simplified to show how current distributes and voltage output is developed. There are four poles and four paths. Each path carries one-fourth of the total armature current (I_T). A voltage (V_P) is generated by each path. The total power (P_T) rating of the armature is $V_P \times I_T$.

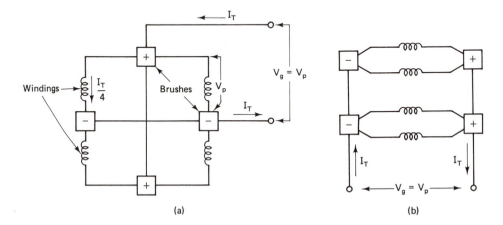

Figure 2-19   Lap-wound dc generator armature circuit: (a) simplified armature circuit; (b) equivalent circuit.

Wave-winding design.   For wave windings, the number of current paths in the armature is determined by the multiplicity of the winding only ($2M$). Current paths are independent of the number of poles. The current flow through each armature coil is equal to the total armature current divided by the number of paths (like lap windings). Wave windings require only two brushes, regardless of the number of poles. However, in large machines more brushes may be used, to reduce the current flow per brush. The example shown in Figure 2-20 is an illustration of a four-pole simplex wave winding. The circuit connections are simplified to show how current distributes and voltage output is developed. There are four poles and two paths. The current flow in each path is equal to $I_T \div 4$. Since there are two paths connected in series, the generated voltage per path is $2 \times V_P$. The power rating is equal to $2 \times V_P \times I_T$, as in lap-wound armatures.

Comparison of lap and wave windings.   Ordinarily, lap windings are used for high-current, low-voltage applications. On the other hand, wave windings lend themselves to low-current, high-voltage applications. It is relatively easy to determine whether an armature is lap-wound or wave-wound. The ends of lap-wound coils are connected to adjacent commutator bars, while wave-wound coils are not. Wave

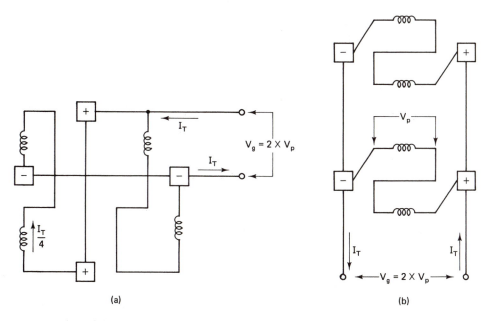

Figure 2-20 Wave-wound dc generator armature circuit: (a) simplified armature circuit; (b) equivalent circuit.

windings require only two brushes, regardless of the number of poles. This characteristic of wave windings is important in applications where brush maintenance and replacement is a serious problem.

REVIEW

2.1. What is the basic difference between a generator and a motor?

2.2. Define the following parts of rotating machines: **(a)** stator, **(b)** rotor, **(c)** yoke, **(d)** field poles, and **(e)** field windings.

2.3. What is the difference between concentrated windings and distributive windings?

2.4. What two basic types of rotors are commonly used in rotating electrical machines?

2.5. Discuss the construction and uses of slip rings and split-ring commutators.

2.6. What is the purpose of brushes used with rotating electrical machines?

2.7. Discuss the construction of the electrical machine shown in Figure 2-7.

2.8. What is the purpose of the field coils of a rotating electrical machine?

2.9. Discuss the method used to connect four field coils in series to produce N-S-N-S polarities.

2.10. What is the purpose of using more than two poles in the construction of a rotating electrical machine?

2.11. Discuss lap-wound and wave-wound armatures.

2.12. How are the voltage current and power ratings of an armature determined?

2.13. Define the following terms used in winding design: **(a)** full-pitch coil, **(b)** fractional-pitch coil, **(c)** chorded windings, **(d)** simplex windings, **(e)** multiplicity, **(f)** pitch factor, and **(g)** triplex windings.

2.14. How is the generated voltage (V_g) of a dc generator calculated?

2.15. How is the number of current paths for a lap-wound armature determined? For a wave-wound armature?

PROBLEMS

2.1. A dc generator has 8×10^5 flux lines per pole, two turns per coil, and rotates at a speed of 1000 r/min. What is the average voltage induced per coil?

2.2. A six-pole dc generator has a triplex lap-wound armature. The magnetic flux per pole is 6×10^4 lines, the speed of rotation is 500 r/min, and there are 250 coils, each having two turns of wire. Calculate the generated voltage.

2.3. A dc generator has 6×10^4 magnetic flux lines per pole, two turns per coil, and a speed of 2000 r/min. What is the V_{avg} per coil?

2.4. A duplex wave-wound armature is used in a four-pole dc generator. The magnetic field flux per pole is 2.4×10^5 lines, the speed of rotation is 1000 r/min, and there are 200 coils, each having two turns. What is the generated voltage?

2.5. A dc generator has 5×10^6 flux lines per coil and rotates at a speed of 600 r/min. Calculate the average voltage induced per coil.

2.6. An eight-pole dc generator has a duplex lap-wound armature. The magnetic flux per pole is 5×10^5 lines, the speed of rotation is 1000 r/min, and there are 200 coils, each having two turns of wire. What is the generated voltage?

2.7. A dc generator has 5.5×10^5 flux lines per pole, two turns per coil, and rotates at a speed of 1800 r/min. Calculate the average voltage per coil.

2.8. A triplex lap-wound armature is used in an eight-pole dc generator. The magnetic field flux per pole is 3.5×10^5 lines, the speed of rotation is 600 r/min, and there are 100 coils, each having two turns. Calculate the generated voltage.

THREE

Electrical Power Basics

Electrical power is the energy source for many machines and appliances used today. Electrical power has thus become a necessity of life. An understanding of electrical power basics is fundamental to the study of rotating electrical machines and power systems.

A block diagram of an electrical power system model was shown in Figure 1-19. *Electrical power production* is an important part of the electrical power system model used in this book. Once electrical power is produced, it must be distributed to the location where it is used. *Electrical power distribution* systems transfer electrical power from one location to another. *Electrical power control* systems are probably the most complex of all the parts of the power system model. There are many types of devices and equipment used to control electrical power. *Electrical power conversion* systems, also called *loads,* convert electrical power into some other form of energy. Other forms of energy include heat, light, and mechanical energy. Conversion systems, such as electric motors, are very important when dealing with energy use.

Each of the blocks shown in the electrical power system model (Figure 1-19) represents one part of the electrical power system. It is important to understand each part of the electrical power system as it relates to rotating electrical machines and transformers. By using the system model, a better understanding of electrical power system operation can be gained.

ELECTRICAL POWER PRODUCTION SYSTEMS

There is a tremendous need for electrical power in our country today. To meet the demand for electrical power, power companies use mechanical generators to produce vast quantities of electrical power at power plants. Generating units at these power plants are the *source* of an electrical power system.

Electrical power is produced in many ways, such as from chemical reactions, heat, light, or mechanical energy. Most electrical power is produced by large generators at power plants. Most plants convert the energy produced by burning coal, oil, or natural gas into electrical energy. Some power plants produce electrical power from the force of flowing water or from nuclear reactions. Generators at power plants are often driven by steam or gas turbines, while hydraulic turbines are used to rotate generators at hydroelectric power plants. Several types of modern electrical power production systems are shown in Figure 3-1.

Various other methods may be used in the future for power production. These alternative *methods* include solar cells, geothermal systems, wind-powered systems,

Figure 3-1 Types of modern electrical power production systems: (a) fossil-fuel system; (b) electrical power plant—fossil-fuel system; (c) hydroelectric system; (d) hydroelectric generator units; (e) hydroelectric system; (f) nuclear-fission power system; (g) nuclear-fission electrical power plant. [(b), Courtesy of Tennessee Valley Authority; (d), (e), courtesy of U.S. Army Corps of Engineers; (g), courtesy of General Electric Company.]

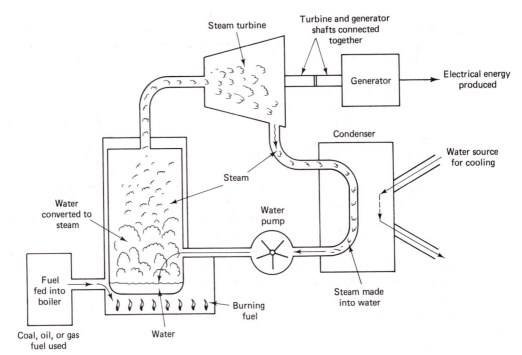

(a)

(b)

(c)

Figure 3-1 (*cont.*)

(d)

(e)

Figure 3-1 (*cont.*)

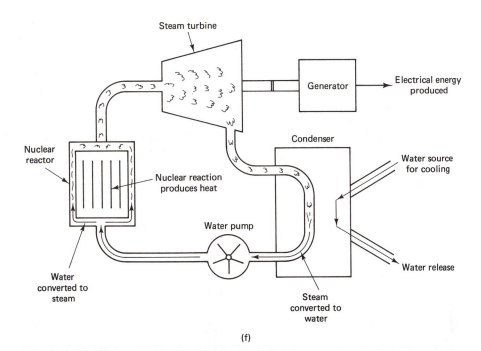

(f)

(g)

Figure 3-1 (*cont.*)

magnetohydrodynamic (MHD) systems, nuclear-fusion systems, and fuel cells. Many of these methods are now being researched as future power sources. Some alternative power production methods are shown in Figure 3-2.

Electrical power is ordinarily produced at power plants that are fossil-fuel steam plants, nuclear-fission steam plants, or hydroelectric plants. Fossil-fuel and nuclear-fission plants use steam turbines to produce mechanical energy. This mechanical energy is used to rotate large three-phase alternators. Hydroelectric plants ordinarily use hydraulic turbines to convert the force of flowing water into mechanical energy which rotates three-phase generators. The types of generators used to produce electrical power are discussed in Chapters 4 and 5.

The supply and demand situation for electrical power is very different from that for most other products. Most products are produced and then sold to consumers at a later time. However, electrical power must be supplied at the same time that it is

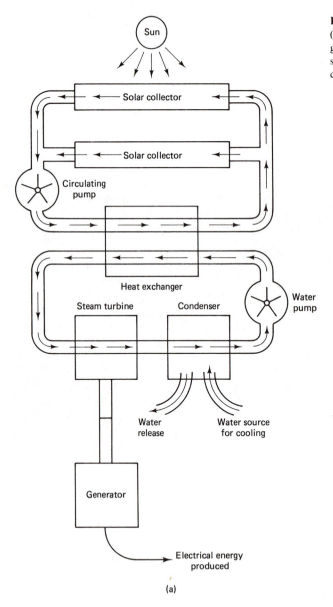

Figure 3-2 Alternative power systems: (a) solar electrical energy system; (b) geothermal power system; (c) wind-energy system; (d) MHD energy system; (e) fuel cell.

(a)

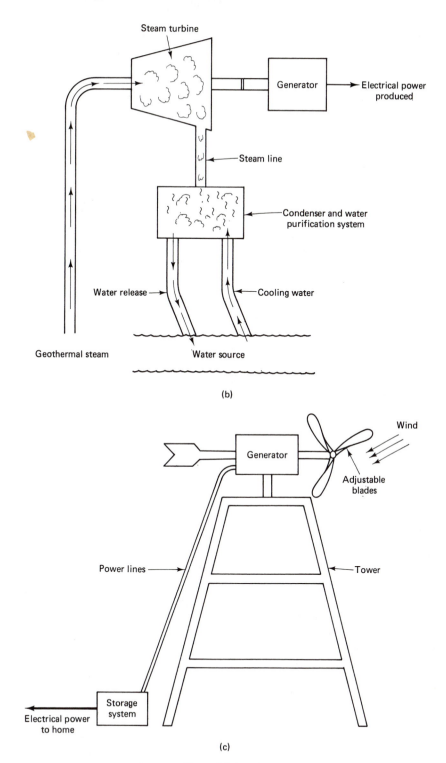

(b)

(c)

Figure 3-2 (*cont.*)

43

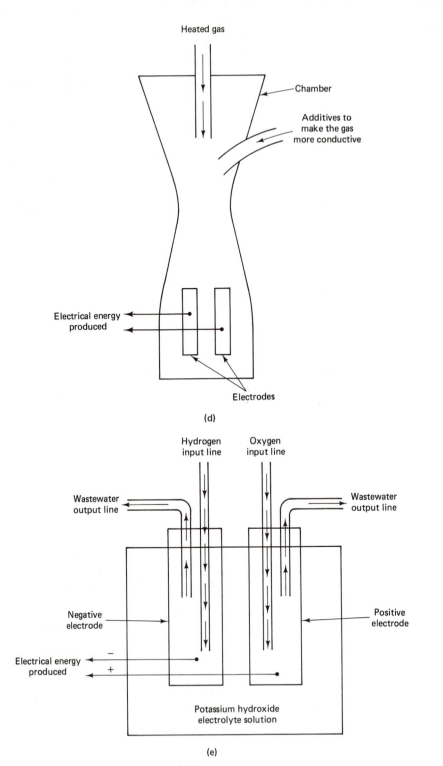

(d)

(e)

Figure 3-2 (*cont.*)

demanded by consumers. There is no simple storage system that may be used to supply additional electrical energy at times of high demand. Enough electrical power must be produced to meet the demand of the consumers at any time. Power companies must predict the electrical power output needed for a particular time of the year, week, or day.

POWER DISTRIBUTION SYSTEMS

The electrical power system delivers power from the source to the load that is connected to it. Power distribution systems are used to provide a means of delivering electrical power from a generating station to the location where it is used. The distribution of electrical power from a power plant is shown in Figure 3-3.

The distribution of electrical power requires many power lines to carry the electrical current from where it is produced to where it is used. Parts of power distribution systems include transformers, conductors, insulators, fuses, circuit breakers, and grounding systems. An electrical distribution substation is shown in Figure 3-4.

Electrical Conductors

Conductors, such as those shown in Figure 3-5, are used to distribute electrical current from one location to another. The unit of measurement for conductors is the circular mil (cmil) since most conductors are round. One mil is equal to a circle whose diameter is 0.001 in. The cross-sectional area of a conductor (in circular mils) is equal to its diameter (D) in mils squared (cmil = D^2). For example, if a conductor is $\frac{1}{8}$ in. in diameter, its circular-mil area is found as follows:

1. $\frac{1}{8}$ in. written as a decimal is 0.125.
2. 0.125 equals 125 mils.
3. Area = D^2 (in mils)
   = $(125)^2$
   = 15,625 cmils

If the conductor is not round in shape, its area is found by another formula. For example, a rectangular copper conductor is 0.25×0.3 in. in size. Its area is found as follows:

$$\text{area} = \frac{0.2 \text{ in.} \times 0.3 \text{ in.}}{0.7854}$$

$$= \frac{200 \text{ mils} \times 300 \text{ mils}}{0.7854}$$

$$= \frac{60,000 \text{ mils}^2}{0.7854}$$

$$= 76,394 \text{ cmils}$$

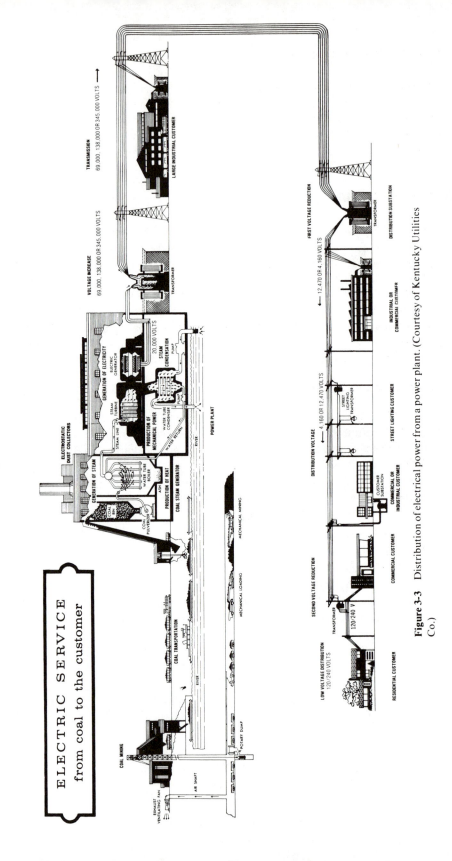

Figure 3-3 Distribution of electrical power from a power plant. (Courtesy of Kentucky Utilities Co.)

Figure 3-4 Electrical substation. (Courtesy of Lapp Insulator, Interpace Corp.)

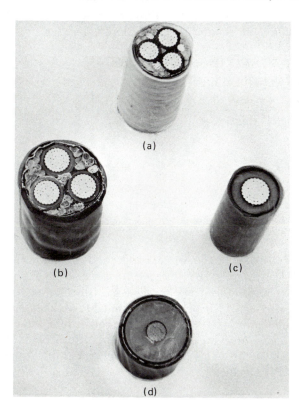

(a)

(b)

(c)

(d)

Figure 3-5 Electrical conductors: (a) three-phase aluminum conductor group, with grounds; (b) three-phase aluminum conductors; (c) aluminum high-current conductor; (d) copper 69-kV conductor.

The resistance of a conductor expresses the amount of opposition it offers to the flow of electrical current. The unit of measurement for resistance is the ohm (Ω). *The resistivity (ρ)* of a conductor is the resistance of a given cross-sectional area and length (*l*) in circular mil-feet (cmil-ft). The resistivity of a conductor changes with the temperature. The resistivities for some common types of conductors are listed in Table 3-1.

TABLE 3-1 RESISTIVITY
OF CONDUCTORS

Conductor	Resistivity at 20° C (Ω/cmil-ft)
Silver	9.8
Copper	10.4
Aluminum	17.0
Tungsten	33.0

The values of Table 3-1 are used to calculate the resistance of any conductor. Resistance increases as the length of the conductor increases. Resistance decreases as the cross-sectional area of a conductor increases. The following method is used to find the resistance of 200 ft of aluminum conductor that is $\frac{1}{8}$ in. in diameter.

1. Aluminum has a resistivity of 17 Ω/cmil-ft (see Table 3-1).
2. Diameter (*D*) equals $\frac{1}{8}$ in. = 0.125 in. = 125 mils.
3. Resistance = $\dfrac{\text{resistivity} \times \text{length (ft)}}{\text{diameter}^2 \text{ (mils)}}$ or $R = \dfrac{\rho \times l}{D^2}$

   $= \dfrac{17 \times 200}{(125^2)}$

   $= \dfrac{3400}{15,625}$

   $= 0.22\ \Omega$

Conductor sizes. Table 3-2 lists the sizes of copper and aluminum electrical conductors. The American Wire Gage (AWG) is used to measure the diameter of conductors. The sizes of conductors range from No. 40 AWG (smallest) to No. 0000 AWG. Larger conductors are measured in thousand circular mil (MCM) units. Notice in Table 3-2 that as the AWG size numbers become smaller, the conductor is larger. Sizes up to No. 8 AWG are solid conductors. Larger wires have from 7 to 61 strands of wire. Table 3-2 also lists the resistance (in ohms per 1000 ft) of copper and aluminum conductors. Notice that smaller conductors have higher resistance.

TABLE 3-2 SIZES OF COPPER AND ALUMINUM CONDUCTORS

Size (AWG or MCM)	Area (cmil)	Number of wires	Diameter of each wire (in.)	Dc resistance at 25° C (Ω/1000 ft)	
				Copper	Aluminum
18	1,620	1	0.0403	6.51	10.7
16	2,580	1	0.0508	4.10	6.72
14	4,110	1	0.0641	2.57	4.22
12	6,530	1	0.0808	1.62	2.66
10	10,380	1	0.1019	1.018	1.67
8	16,510	1	0.1285	0.6404	1.05
6	26,240	7	0.0612	0.410	0.674
4	41,740	7	0.0772	0.259	0.424
3	52,620	7	0.0867	0.205	0.336
2	66,360	7	0.0974	0.162	0.266
1	83,690	19	0.0664	0.129	0.211
0	105,600	19	0.0745	0.102	0.168
00	133,100	19	0.0837	0.0811	0.133
000	167,800	19	0.0940	0.0642	0.105
0000	211,600	19	0.1055	0.0509	0.0836
250	250,000	37	0.0822	0.0431	0.0708
300	300,000	37	0.0900	0.0360	0.0590
350	350,000	37	0.0973	0.0308	0.0505
400	400,000	37	0.1040	0.0270	0.0442
500	500,000	37	0.1162	0.0216	0.0354
600	600,000	61	0.0992	0.0180	0.0295
700	700,000	61	0.1071	0.0154	0.0253
750	750,000	61	0.1109	0.0144	0.0236
800	800,000	61	0.1145	0.0135	0.0221
900	900,000	61	0.1215	0.0120	0.0197
1000	1,000,000	61	0.1280	0.0108	0.0177

AWG sizes brackets rows 18 through 0000.
MCM sizes brackets rows 250 through 1000.

Ampacity of conductors. The amount of current a conductor can carry is called its ampacity. All metal materials will conduct electrical current; however, copper and aluminum are the two most widely used conductors. Copper is used more frequently since it is the better conductor of the two and is physically stronger. Aluminum is used where weight is a factor, such as for long-distance outdoor power distribution lines. The weight of copper is almost three times that of the same amount of aluminum. The ampacity of an aluminum conductor is less than that of the same-size copper conductor.

The ampacity of conductors depends on several factors. The types of material, cross-sectional area, and type of location in which they are installed affect conductor ampacity. Conductors in the open or in "free air" transfer heat much more rapidly than when enclosed in a metal raceway or plastic cable. When several conductors are in the same enclosure, heat transfer becomes a problem.

Table 3-3 is used to find conductor ampacity for electrical wiring design. Tables like this one are given in the *National Electrical Code®* (NEC), which is used to provide

TABLE 3-3 CONDUCTOR AMPACITY

		Copper		Aluminum	
	Wire Size	With R, T, TW insulation	With RH, RHW, TH, THW insulation	With R, T, TW insulation	With RH, RHW, TH, THW insulation
AWG sizes	14	15	15		
	12	20	20	15	15
	10	30	30	25	25
	8	40	45	30	40
	6	55	65	40	50
	4	70	85	55	65
	3	80	100	65	75
	2	95	115	75	90
	1	110	130	85	100
	0	125	150	100	120
	00	145	175	115	135
	000	165	200	130	155
	0000	195	230	155	180
MCM sizes	250	215	255	170	205
	300	240	285	190	230
	350	260	310	210	250
	400	280	335	225	270
	500	320	380	260	310
	600	355	420	285	340
	700	385	460	310	375
	750	400	475	320	385
	800	410	490	330	395
	900	435	520	355	425
	1000	455	545	375	445

standards for electrical wiring design. Table 3-3 is used to find the ampacity of conductors when not more than three are mounted in a raceway (enclosure) or a cable.

As an example, find the ampacity of a No. 2 copper conductor with RHW insulation. (1) Looking on the left column at the No. 2 wire size, (2) find 115 A along the horizontal row under the RHW column. Thus No. 2 copper conductors with RHW insulation will carry 115 A (maximum) of current.

Insulation

Synthetic insulation for electrical conductors is classified into two broad categories: thermosetting and thermoplastic. The mixtures of materials makes the number of insulations available almost unlimited. Most insulation is made of synthetic materials combined to provide the necessary physical and electrical characteristics. The operating conditions in which conductors are used determine the type of insulation required. Insulation must withstand the heat of the surrounding atmosphere and the

TABLE 3-4 TYPES OF INSULATION

Abbreviation	Type of insulation
R	Rubber—140° F
RH	Heat-resistant rubber—167° F
RHH	Heat-resistant rubber—194° F
RHW	Moisture and heat-resistant rubber—167° F
T	Thermoplastic—140° F
THW	Moisture and heat-resistant thermoplastic—167° F
THWN	Moisture and heat-resistant thermoplastic with nylon—194° F

heat produced by current flow through the conductor. Large currents produce excessive conductor heat, which can cause insulation to melt or burn. This is why overcurrent protection (fuses or circuit breakers) is required as a safety factor to prevent fires. The ampacity of a conductor is affected by the type of insulation used. The NEC has developed a system of abbreviations for identifying types of insulation. Some of the abbreviations are listed in Table 3-4.

Fuses and Circuit Breakers

There are many devices used to protect electrical power systems from damage, including switches, fuses, circuit breakers, lightning arresters, and protective relays. Some of these devices automatically disconnect electrical equipment before any damage can occur due to overcurrents, while others sense changes from the normal operation of equipment. The most common electrical problems that require protection are "short-circuit" conditions and other problems, such as overvoltage, undervoltage, and changes in frequency. The purpose of any type of protective device is to stop current flow in a system when excess current flows through the system.

One type of protective device is a fuse, examples of which are shown in Figure 3-6. Fuses ordinarily are low-cost items compared to other protective devices. However, a disadvantage of fuses is that replacements are required when they become inoperative due to excessive current flow.

A circuit breaker is another type of overload device; it functions in the same way as a fuse to prevent overcurrents in systems. Circuit breakers are usually made in molded plastic cases which mount in metal power distribution panels (see Figure 3-7). Circuit breakers are designed so that they will automatically open a circuit when a rated current value is exceeded. For instance, a 20-A circuit breaker will open when more than 20 A of current flows in a circuit and it may be reset manually. Most circuit breakers use a thermal tripping element or a magnetic trip element which allows them to sense current flow and respond to overcurrents.

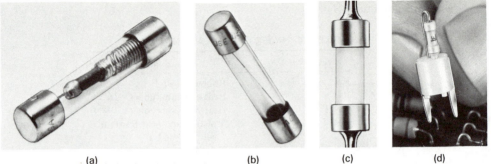

(e)

Figure 3-6   Types of fuses: (a) slow-blow fuse element; (b) regular fuse element; (c) in-line fuse; (d) miniature plug-in fuse for printed circuit boards; (e) "socket" fuses. (Courtesy of Littlefuse, Inc.)

Figure 3-7   Circuit breaker construction. (Courtesy of Heinemann Electric Co.)

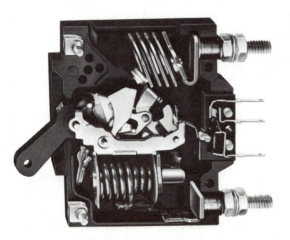

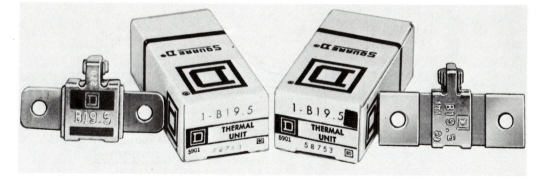

Figure 3-8 Thermal motor overload devices. (Courtesy of Square-D Co.)

Motors are an example of equipment that must be protected from excessive overheating. This protection is provided by magnetic or thermal protective devices such as those shown in Figure 3-8. These devices may be mounted inside the motor or in the equipment used to start and stop the motor. When a motor overheats, the protective device is used to disconnect the motor automatically from its power supply.

POWER DISTRIBUTION IN BUILDINGS

Electrical power is distributed over long-distance power lines to the location where it is to be used. It is then distributed inside a building by an interior power distribution system. Parts of power distribution systems in buildings include conductors, feeder systems, branch circuits, grounding systems, and protective equipment.

Most electrical distribution inside buildings is through wires and cables contained in *raceways,* which are plastic or metal enclosures used to hold conductors, which transfer power to the electrical equipment used in a building. Copper conductors are ordinarily used for interior electrical power distribution. The size of each conductor depends on the current rating of the circuit in which it is used.

Feeder lines and branch circuits. The conductors that carry electrical current throughout a building are called *feeder lines* and *branch circuits.* Feeder lines supply electrical power to the branch circuits which are connected to them and may be mounted either overhead or underground. Underground systems cost more, but they are much more attractive in appearance. Power distribution is from feeder lines, through protective equipment, to branch circuits inside a building. Each branch circuit has a fuse or circuit breaker connected in series. The relationship of feeders and branch circuits is shown in Figure 3-9.

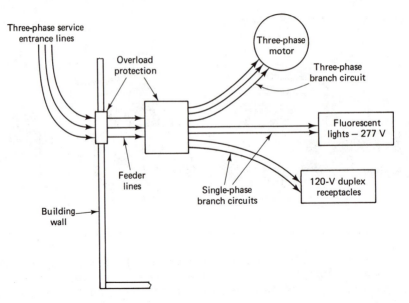

Figure 3-9 Relationship of feeders and branch circuits.

Electrical service entrances and distribution. Electrical power is brought from overhead power lines or from underground cables into a building through a *service entrance,* as shown in Figure 3-10. A meter used to measure power (kilowatt-hour meter), such as the one shown in Figure 3-11, is also part of the service

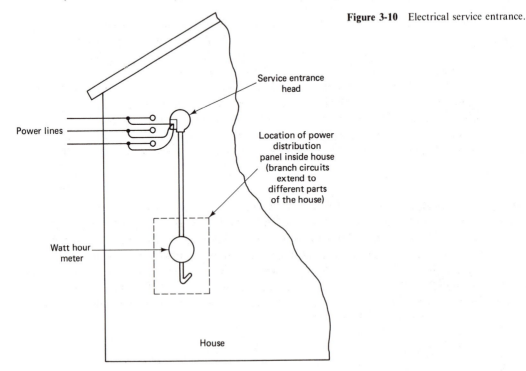

Figure 3-10 Electrical service entrance.

Figure 3-11 Kilowatthour meter. (Courtesy of Sangamo-Schlumberger.)

entrance. It is necessary to *ground* the power distribution system at the service entrance location. This is done by a "grounding electrode," which is a metal rod driven deep into the ground so that a grounding conductor may be attached securely to the grounding electrode. The grounding conductor must then be connected to all neutral conductors and safety ground conductors of the system.

Electrical power lines are brought into residential buildings as a three-wire 120/240-V service entrance. The service entrance consists of the three wires which are inside a large metal conduit along the outside wall of the building. These large wires are brought into the home and attached to a power distribution panel. The local electric company installs the service entrance, including the watthour meter, which measures electrical power consumption in the building.

The service entrance wires are securely connected into the power distribution panel. This metal enclosure has fuses or, more commonly, circuit breakers which provide overload protection for the entire building. Several branch circuits may be wired from the power distribution panel, each with its own protective fuse or circuit breaker. Overload protection either by fuses or circuit breakers is important to prevent possible fires caused by the heating effect of excessive electrical current flowing through conductors. Fuses or circuit breakers are placed in series with all hot conductors of the branch circuits in buildings.

Power distribution panels are rated according to the amount of electrical energy they will distribute to a building. A common panel used for residential buildings is rated at 200 A, with voltages of 120 and 240 V. Larger distribution panels may be used for industrial and commercial buildings. Two types of distribution panels are shown in Figure 3-12. Distribution systems are discussed in detail in Chapter 6.

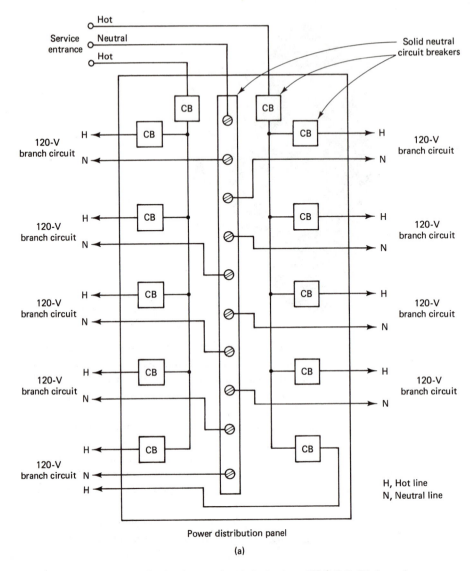

Power distribution panel

(a)

Figure 3-12 Power distribution panels: (a) single-phase 120/240 V; (b) three-phase 120/208 V.

Electrical wiring. A common electrical wire used in residential buildings is called nonmetallic sheathed cable or Romex. One type of cable is No. 12-3 WG. The No. 12 means that the size of the wires is No. 12 as measured with an American Wire Gage. The "3" means that there are three current-carrying wires inside the cable, and the "WG" means "with ground." There is an additional wire which serves as a safety ground to prevent electrical shock.

The insulation color on the wires of nonmetallic cable is very important. Hot

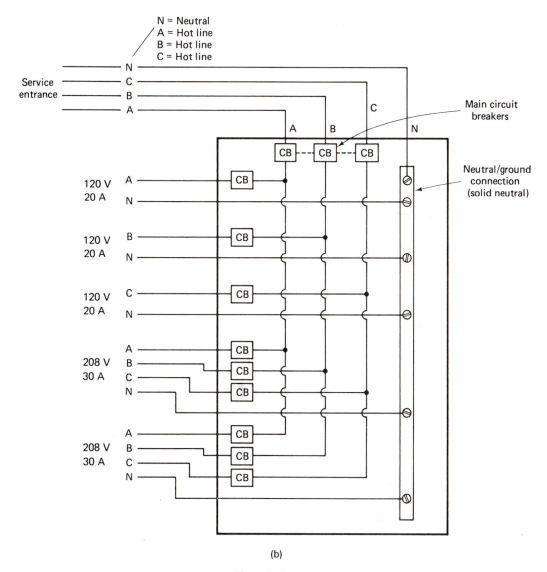

(b)

Figure 3-12 (*cont.*)

wires have either black or red insulation, while neutral wires have white insulation. The safety ground ordinarily has either green insulation or is a bare wire. The neutral and safety ground wires are always connected to the neutral-ground strip of the power distribution panel (see Figure 3-12).

Another example of electrical cable is No. 12-2 WG cable. This type has wires that have a diameter of No. 12 size with one hot conductor and one neutral conductor. The "WG" indicates that the cable has a safety ground wire. Type No. 12-2 WG cable is used for many 120-V branch circuits in residential buildings. Smaller numbers indicate larger conductors. For instance, a No. 8 wire is larger than a No. 12 wire and a No. 14 wire is smaller than a No. 12 wire.

ELECTRICAL POWER CONTROL

Control is the most complex part of the electrical power system. Control equipment and devices are used with many types of electrical loads. Equipment such as motors, lighting, and heating systems require control systems. Control systems used for electrical equipment are discussed in Chapter 12.

ELECTRICAL POWER CONVERSION (LOADS)

Electrical loads are also important parts of electrical power systems since they convert energy from one form to another. An electrical load converts electrical energy to some other form, such as heat, light, or mechanical energy. Electrical loads are usually classified by the function they perform, such as lighting, heating, or mechanical.

Load characteristics. The types of power supplies and distribution systems used with electrical power systems are determined by the characteristics of the load. Electrical loads may be either resistive, inductive, capacitive, or a combination of these. Each type of load has different effects on a power system. The nature of alternating-current power causes certain electrical properties to exist.

A major characteristic that affects electrical power system operation is the presence of inductive loads. Since electrical motors have windings, they are a major type of inductive load. To counteract the inductive effects, capacitors may be used as part of the power system. Capacitor units are often located at substations to improve the *power factor* of the system. The inductive effect reduces the actual amount of energy that is converted to another form by an electrical load.

Demand factor. Demand factor is a ratio used to express the difference between the average power used by a building and the peak power used. The formula used to calculate demand factor is:

$$\text{demand factor} = \frac{\text{average demand (kW)}}{\text{peak demand (kW)}}$$

For example, if the average electrical power demand for a building is 2000 kW and the peak demand is 2950 kW, the demand factor is

$$\frac{2200 \text{ kW}}{2950 \text{ kW}} = 0.745$$

The average demand for a building is the average electrical power used over a period of time. The peak demand is the maximum amount of power used during the same time

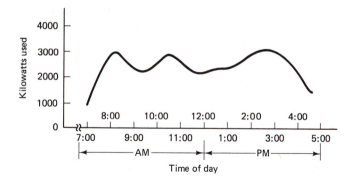

Figure 3-13 Industrial load profile.

period. The load profile shown in Figure 3-13 shows how a typical industrial power demand varies during a working day. A peak demand that is far higher than average demand causes a poor demand factor. Low demand factors cause high electric bills for industries and commercial buildings. Utility companies must design power distribution systems and provide generating capacity which is capable of meeting this peak demand.

Power-factor correction.   Most industries use many electric motors (inductive loads), which cause the power system to operate at a power factor of less than 1. Power systems for industries should not operate at low power factors. This situation would cause the electrical power system to have much greater capacity than is actually needed.

A specific value of voltamperes (voltage × current or VA) is supplied to industries by electrical power systems. If the power factor (PF) of the industry's system is low, the current draw from the system is higher than necessary. The power converted by the total industrial power system is equal to VA × PF. Power factor decreases as reactive power (unused power associated with motor use) increases. Thus, when reactive power draw by electric motors increases, more voltamperes (VA) are drawn from the power system. The voltage of the system remains constant; thus the current value must increase since $P = V \times A$. An increase in reactive power draw by motors causes a decrease in the power factor. Power factor is found as follows:

$$PF = \frac{\text{true power (W)}}{\text{apparent power (VA)}}$$

For example, if the true power converted by an electric motor is 5000 W and it draws 25 A of current from a 240-V power line, its power factor is

$$PF = \frac{W}{VA} = \frac{5000 \text{ VA}}{240 \text{ V} \times 25 \text{ A}} = 0.833 \quad \text{or} \quad 83.3\%$$

Industries should attempt to improve power factor to avoid high energy costs. A high power factor (close to 1.0) causes more economical use of electrical energy. Two methods are used to increase power factor: (1) power factor-corrective capacitors and (2) three-phase synchronous capacitors (motors). The effect of capacitance in ac systems is the opposite of the effect of inductance; thus capacitive and inductive effects counteract one another. Either of these two methods may be used to add the effect of capacitance to an ac power system to cause the power factor to increase.

REVIEW

3.1. What are the five parts of an electrical power system, and what is the primary purpose of each of these parts?

3.2. Describe the major methods used to produce electrical power.

3.3. What are some possible methods of producing electrical power in the future?

3.4. Discuss the operation of the power distribution system shown in Figure 3-3.

3.5. What is a circular mil?

3.6. How may the resistance of a conductor be found?

3.7. Discuss the advantages and disadvantages of copper and aluminum conductors.

3.8. What is the difference between a feeder line and a branch circuit?

3.9. What is the principal difference between a fuse and a circuit breaker?

3.10. Describe an electrical service entrance for a residential building.

3.11. What is power factor?

3.12. What is demand factor?

3.13. What is power factor correction, and how may this process be accomplished?

PROBLEMS

3.1. Convert the following diameters of electrical conductors into circular mils.
   (a) 0.65 in.   **(b)** 0.32 in.
   (c) 0.3 in.   **(d)** $\frac{1}{2}$ in.
   (e) $\frac{1}{4}$ in.   **(f)** $\frac{5}{16}$ in.
   (g) $\frac{1}{8}$ in.   **(h)** $\frac{3}{4}$ in.
   (i) 0.91 in.   **(j)** 0.8 in.

3.2. Calculate the cross-sectional area of the following electrical conductors.
   (a) $\frac{3}{4}$ inch $\times$ $\frac{3}{8}$ in.   **(b)** 0.33-in. diameter
   (c) $\frac{3}{16}$-in. diameter   **(d)** 0.4-in. square
   (e) $\frac{1}{2}$ inch $\times$ $\frac{5}{8}$ in.

3.3. Calculate the resistance of the following electrical conductors.
   (a) 250 ft of No. 10 aluminum   **(b)** 1500 ft of No. 8 aluminum
   (c) 280 ft of 500 MCM copper   **(d)** 3100 ft of No. 00 copper
   (e) 350 ft of No. 12 copper   **(f)** 520 ft of No. 1 aluminum
   (g) 1850 ft of No. 6 copper

3.4. Determine from Table 3-3 the minimum sizes of conductors needed to carry the following currents.

(a) 85 A using TH aluminum
(b) 520 A using T copper
(c) 680 A using RHW copper
(d) 220 A using R copper
(e) 290 A using TW copper
(f) 150 A using RH copper
(g) 92 A using THW aluminum
(h) 360 A using THW copper

3.5. Find the minimum sizes of THW aluminum conductors to replace the following copper conductors.

(a) No. 8, RHW
(b) No. 4, T
(c) No. 00, R
(d) 650 MCM, TH
(e) 500 MCM, TW
(f) 200 MCM, RHW

3.6. Calculate the demand factor of a commercial building that has an average electrical power usage of 3520 kW and a peak power usage of 4150 kW.

3.7. Calculate the power factor of an electrical system that has a true power consumption of 3800 kW. The system voltage is 120 V and the current is 40 A.

3.8. Calculate the power factor of a 480-V ac motor that converts 5500 W of power and draws 30 A of current from its power source.

FOUR

Alternating-Current Generators (Alternators)

The vast majority of the electrical power used in the United States is alternating current (ac) produced by mechanical generators at power plants. These power plants, located throughout our country, provide the necessary electrical power to supply our homes and industries. Most generators or *alternators* produce three-phase alternating current; however, single-phase generators are also used for certain applications with smaller power requirements. The operation of all mechanical generators relies on the fundamental principle of *electromagnetic induction* which was discussed in Chapter 1. Ac generators are commonly called *alternators,* since they produce alternating-current electrical power.

This chapter provides an overview of the basic characteristics and types of ac generators. Preceding the material dealing specifically with ac generators, a brief review of ac circuits and power relationships in ac circuits is presented.

REVIEW OF AC CIRCUITS

Alternating-current (ac) circuits are ordinarily more complex than direct-current (dc) circuits. Alternating-current circuits have a *source* of energy and a *load* in which power conversion takes place. There are three important characteristics present in ac circuits: (1) resistance, (2) inductance, and (3) capacitance. Also, there are two common types of ac voltage: (1) single-phase and (2) three-phase.

Phase relationships in ac circuits. The term *phase* refers to time or the angular difference between one point and another. If two ac sine-wave voltages reach their zero and maximum values at the same time, they are considered to be *in phase*.

Figure 4-1 shows two ac voltages that are in phase. If two voltages reach their zero and maximum values at different times, they are *out of phase*. Figure 4-2 shows two ac voltages that are out of phase. Phase difference is given in degrees; thus the voltages shown are out of phase an angle of 90°. Phase relationships are affected by the amount of resistance, inductance, and capacitance in ac circuits.

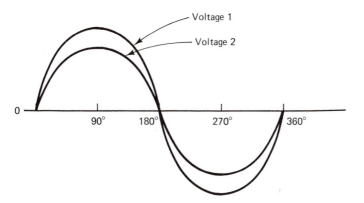

Figure 4-1   Two ac voltages that are in phase.

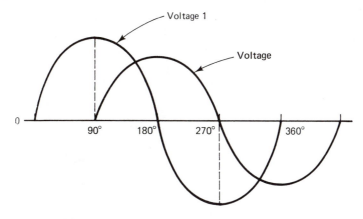

Figure 4-2   Two ac voltages that are out of phase by an angle of 90°.

Resistive circuits.   The simplest type of electrical circuit is a resistive circuit, as illustrated in Figure 4-3(a). A resistive circuit offers the same type of opposition to ac as it does to dc. In dc circuits, the following Ohm's law relationships exist:

$$V = I \times R$$

$$I = \frac{V}{R}$$

$$R = \frac{V}{I}$$

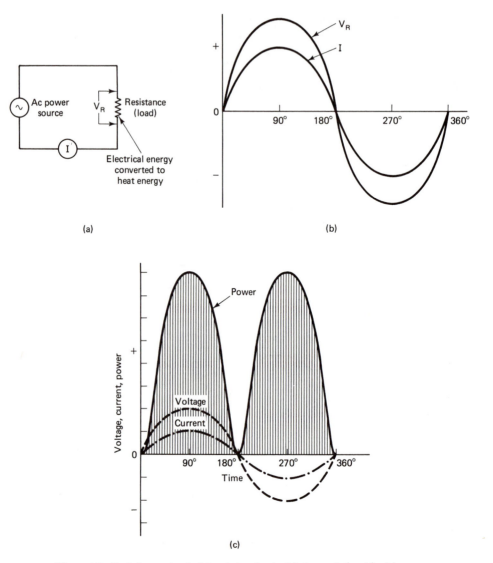

(a)

(b)

(c)

Figure 4-3   Resistive ac circuit: (a) resistive circuit; (b) phase relationship; (c) power curves.

where V = voltage in volts
      I = current in amperes
     R = resistance in ohms

    By looking at the waveforms of Figure 4-3(b), it can be seen that the voltage (V) and current (I) in a purely resistive circuit are in phase. An in phase relationship exists when the minimum and maximum values of both voltage and current occur at the same time. The power converted by the resistance is a product of voltage times current ($P=$

$V \times I$). The power curve is shown in Figure 4-3(c). When an ac circuit contains only resistance, its behavior is very similar to that of a dc circuit.

Inductive circuits. The property of inductance (L) is common in ac circuits. Inductance adds more complexity to the relationship between voltage and current in ac circuits. Motors, generators, and transformers are common ac circuits which have inductance. Inductance is due to the *counterelectromotive force* (CEMF) produced when a magnetic field is developed around a coil of wire. The magnetic field produced around coils affects circuit action. The CEMF produced by a magnetic field offers opposition to *change* in the current flow of a circuit. In an inductive circuit, voltage *leads* the current. If the circuit were purely inductive (containing no resistance), the voltage would lead the current by 90°, as shown in the inductive circuit of Figure 4-4.

A purely inductive circuit does not convert any power in the load. All power is delivered back to the source. Refer to points A and B on the waveforms of Figure 4-4(c). At the peak of each waveform the value of the other waveform is zero. The power curves are equal and opposite, and thus cancel each other out. Where voltage and current values are positive, the power is also positive, since the product of two positive values is positive. When voltage is positive and current is negative, the product of the two is negative and the power converted is also negative. Negative power means that electrical energy is returned from the load to the source without being converted to another form. The power converted in a purely inductive circuit is equal to zero.

The opposition to the flow of ac current is an inductive circuit is due to the property of *inductance* (L). The amount of opposition to current flow of an inductive circuit depends on the resistance of the wire and the magnetic properties of the inductance. The opposition due to the inductance or magnetic effect is called *inductive reactance* (X_L). X_L varies with the applied frequency and is found by using the formula

$$X_L = 2\pi \times f \times L$$

where $2\pi = 6.28$
f = applied frequency in hertz
L = inductance in *henrys*

The basic unit of inductance is the henry.

At zero frequency (or dc), there is no opposition due to inductance. Only a coil's resistance limits current flow. As frequency applied to a circuit increases, the inductive effect becomes greater. The inductive reactance of an ac circuit usually has more effect on current flow than does the resistance. An ohmmeter measures *dc resistance* only. Inductive reactance must be calculated or determined experimentally.

Ohm's law ordinarily is not applied in the same way to inductive ac circuits as it is with dc circuits. Ohm's law calculations must consider both the resistance (R) and the inductive reactance (X_L), since both values limit current flow. The total opposition to current flow is called *impedance* (Z). Impedance is a combination of resistance and reactance in ac circuits. In ac circuits with resistance and reactance, the Ohm's law relationship is $I = V/Z$.

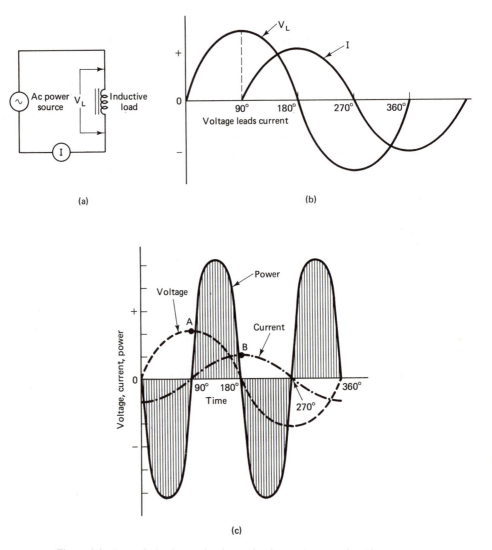

Figure 4-4 Purely inductive ac circuit: (a) circuit; (b) phase relationship; (c) power curves.

Capacitive circuits. Figure 4-5 shows a capacitive circuit. Capacitors are devices that have the ability to store electrical charge and have many applications in electronic circuits. The operation of a capacitor in a circuit depends on its ability to charge and discharge.

If dc voltage is applied to a capacitor, the capacitor will charge to the value of that dc voltage. When the capacitor is fully charged, it blocks the flow of direct current. However, if ac is applied to a capacitor, the changing value of current will cause the capacitor to charge and discharge. The voltage and current waveforms of purely

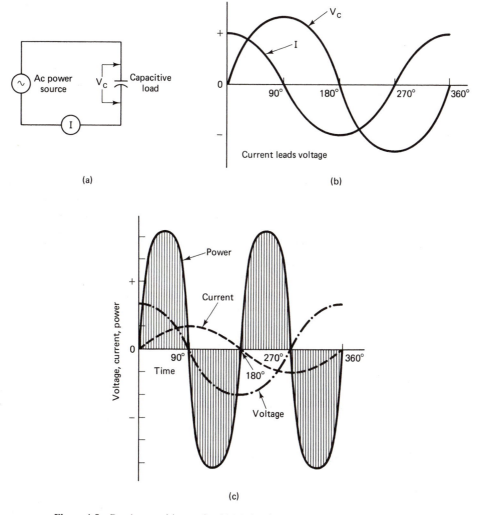

Figure 4-5 Purely capacitive ac circuit: (a) circuit; (b) phase relationships; (c) power curves.

capacitive circuit (no resistance) are shown in Figure 4-5(b). The highest amount of current flows in a capacitive circuit when the voltage changes most rapidly. The most rapid change in voltage occurs at the 0° and 180° positions as polarity changes from (−) to (+). At these positions, maximum current is developed in the circuit. The rate of change of the voltage is slow near the 90° and 270° positions. A small amount of current flows at these positions. Current *leads* voltage by 90° in a purely capacitive circuit. No power is converted in a purely capacitive circuit, just as in the purely inductive circuit. Figure 4-5(c) shows that the positive and negative power waveforms cancel each other out.

Capacitance is the property of a circuit to oppose changes in *voltage*. Capacitance stores energy in an *electrostatic field*. A capacitor is measured by a unit called the *farad*. One farad is the amount of capacitance that permits a current of 1 ampere to flow when the voltage change across the plates of a capacitor is 1 volt per second. The farad is too large for practical use as a unit of measurement. The microfarad (one-millionth of a farad, abbreviated μF) is the most common unit of capacitance. For high-frequency ac circuits, the microfarad is also too large. The unit micro-microfarad (one-millionth of a microfarad) is then used. This unit is ordinarily called the picofarad (pF) to avoid confusion.

Leading and lagging currents in ac circuits.   Inductance and capacitance cause ac circuits to react differently from circuits with resistance only. Inductance opposes changes in *current*, and capacitance opposes changes in *voltage*. In dc circuits and purely resistive ac circuits voltage and current change in step or in phase with each other [see Figure 4-3(b)]. Ac circuits have different effects because the voltage is constantly changing. The current in an inductive circuit lags behind the ac source voltage. Thus voltage *leads* current in an inductive circuit [see Figure 4-4(b)].

The current in a capacitive circuit is maximum when the applied voltage is zero. This is true with both dc and ac circuits. The capacitive current *leads* the applied voltage. The phase relationships of voltage and current in a capacitive circuit are shown in Figure 4-5(b). The 90° phase difference between current and voltage is the maximum that is theoretically possible. In actual ac circuits, it is not possible to have purely inductive or purely capacitive circuits since resistance is always present in actual circuits. Voltage and current in a resistive circuit are always in phase. Combinations of resistance, inductance, and capacitance cause current flow to vary in phase between 90° lagging and 90° leading.

VECTOR OR PHASOR DIAGRAMS

A vector or phasor diagram for resistive, inductive, and capacitive circuits is shown in Figure 4-6. Vector (phasor) diagrams are helpful in ac circuit analysis. Waveforms may be used to show phase relationships; however, it is easier to use phasor diagrams.

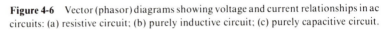

Figure 4-6   Vector (phasor) diagrams showing voltage and current relationships in ac circuits: (a) resistive circuit; (b) purely inductive circuit; (c) purely capacitive circuit.

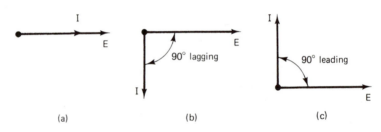

Vectors are straight lines that have specific *direction* and length (*magnitude*) which are used to represent various quantities in ac circuits. The term "phasor" is used to denote the phase difference among ac quantities.

A horizontal line is drawn when beginning a phasor diagram, with its left end as the reference point. In the diagrams of Figure 4-6, the voltage phasor is the horizontal reference line. For the purely inductive circuit, the current phasor is drawn in a *clockwise* direction from the voltage vector. This represents current *lagging* voltage in an inductive circuit. For the purely capacitive circuit, the current phasor is drawn in a *counterclockwise* direction from the voltage phasor. This represents current *leading* voltage in a capacitive circuit.

Resistive-inductive (*RL*) circuits.   All ac circuits have some resistance. The presence of inductance in a circuit causes a condition similar to that shown in Figure 4-7. Voltage leads current by 30° in the phasor diagram. The angular separation between voltage and current is the circuit *phase angle*. The phase angle increases as the inductance of the circuit increases. This type of circuit is called a resistive–inductive (*RL*) circuit. All ac circuits that contain either inductance or capacitance are referred to as *reactive* circuits. This is due to the inductive or capacitive reactance introduced into the circuit.

Figure 4-7   Resistive–inductive (*RL*) circuit: (a) circuit; (b) power curves; (c) vector diagram.

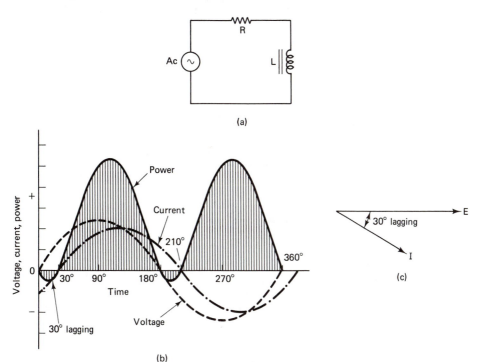

Compare the waveforms of the purely inductive circuit of Figure 4-4 to those of the *RL* circuit of Figure 4-7. In the resistive–inductive (*RL*) circuit, part (but not all) of the power supplied from the source is converted to another form of energy in the load. During the intervals from 0 to 30° and from 180 to 210°, negative power is produced. The remainder of the ac cycle produces positive power.

Resistive-capacitive (*RC*) circuits.   Since all circuits contain some resistance, a more practical capacitive circuit is the (*RC*) circuit shown in Figure 4-8. In an *RC* circuit, current leads voltage by a phase angle between 0 and 90°. Figure 4-8 shows an *RC* circuit in which current leads voltage by 30°. No power is converted in the circuit during the intervals 0 to 30° and 180 to 210°. In the *RC* circuit, most (but not all) of the electrical energy supplied by the source is converted to another form of energy in the load. This circuit should be compared to the purely capacitive circuit of Figure 4-5.

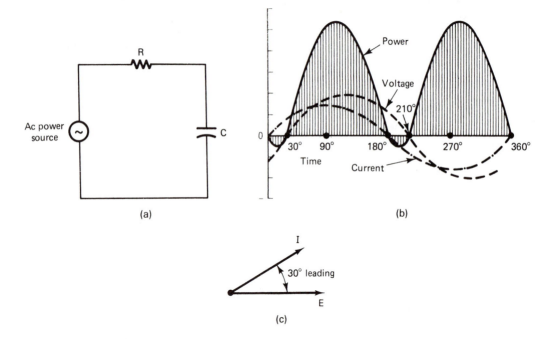

Figure 4-8   Resistive–capacitive (*RC*) circuit: (a) circuit; (b) power curves; (c) vector diagram.

Series ac circuits.   In series ac circuits, the current is the same in all parts of the circuit. The voltage must be added *vectorally* by using a *voltage triangle*. Impedence (*Z*) of a series ac circuit is found by using an *impedance triangle*. Power values are found by using a *power triangle*. These methods must be used due to the phase relationships of ac circuits. Triangles provide an easy way to illustrate ac circuit characteristics.

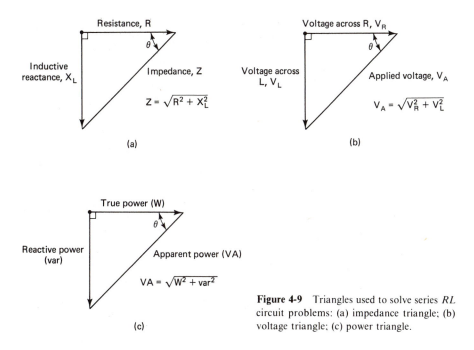

Figure 4-9 Triangles used to solve series *RL* circuit problems: (a) impedance triangle; (b) voltage triangle; (c) power triangle.

Series *RL* circuits.   When an ac voltage is applied to a series *RL* circuit, the current is the same through each component. The voltage drops across each component distribute according to the values of resistance (*R*) and inductive reactance (X_L) in the circuit.

The total opposition to current flow in any ac circuit is called *impedance* (*Z*). Both resistance (*R*) and reactance (*X*) in ac circuits oppose current flow. The impedance of a series *RL* circuit is found by using either of these formulas:

$$Z = \frac{V}{I} \qquad \text{or} \qquad Z = \sqrt{R^2 + X_L^2}$$

The impedance of a series *RL* circuit may also be found by using an *impedance triangle* [see Figure 4-9(a)]. An impedance triangle is formed by the three quantities that oppose alternating current. A triangle may be used to compare voltage drops in series *RL* circuits. Voltage across the inductance (V_L) leads voltage across the resistance (V_R) by 90°. V_A is the voltage applied to the circuit. Since these values form a right triangle, the value of V_A may be found by using the formula $V_A = \sqrt{V_R^2 + V_L^2}$ [see Figure 4-9(b)]. The power triangle for a series *RL* circuit is shown in Figure 4-9(c). It will be explained later. An example of a series *RL* circuit problem is shown in Figure 4-10.

Series *RC* circuits.   Series *RC* circuits are similar to series *RL* circuits; however, in capacitive circuits, current *leads* voltage. Thus the reactive values of *RC* circuits act in opposite directions compared to *RL* circuits. Figure 4–11 shows the

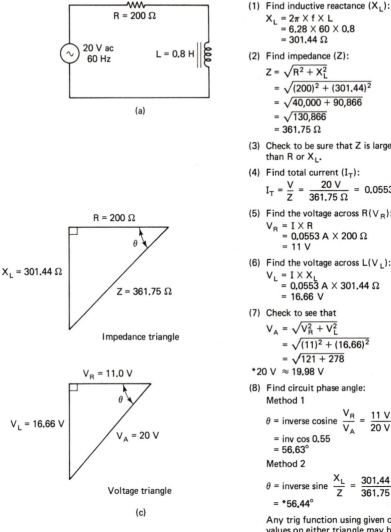

(1) Find inductive reactance (X_L):
$$X_L = 2\pi \times f \times L$$
$$= 6.28 \times 60 \times 0.8$$
$$= 301.44 \ \Omega$$

(2) Find impedance (Z):
$$Z = \sqrt{R^2 + X_L^2}$$
$$= \sqrt{(200)^2 + (301.44)^2}$$
$$= \sqrt{40,000 + 90,866}$$
$$= \sqrt{130,866}$$
$$= 361.75 \ \Omega$$

(3) Check to be sure that Z is larger than R or X_L.

(4) Find total current (I_T):
$$I_T = \frac{V}{Z} = \frac{20 \text{ V}}{361.75 \ \Omega} = 0.0553 \text{ A}$$

(5) Find the voltage across R(V_R):
$$V_R = I \times R$$
$$= 0.0553 \text{ A} \times 200 \ \Omega$$
$$= 11 \text{ V}$$

(6) Find the voltage across L(V_L):
$$V_L = I \times X_L$$
$$= 0.0553 \text{ A} \times 301.44 \ \Omega$$
$$= 16.66 \text{ V}$$

(7) Check to see that
$$V_A = \sqrt{V_R^2 + V_L^2}$$
$$= \sqrt{(11)^2 + (16.66)^2}$$
$$= \sqrt{121 + 278}$$
$$*20 \text{ V} \approx 19.98 \text{ V}$$

(8) Find circuit phase angle:
Method 1
$$\theta = \text{inverse cosine } \frac{V_R}{V_A} = \frac{11 \text{ V}}{20 \text{ V}} = 0.55$$
$$= \text{inv cos } 0.55$$
$$= 56.63°$$
Method 2
$$\theta = \text{inverse sine } \frac{X_L}{Z} = \frac{301.44 \ \Omega}{361.75 \ \Omega} = 0.833$$
$$= *56.44°$$

Any trig function using given or calculated values on either triangle may be used to find phase angle.

(b)

*Slight difference is due to rounding off of numbers as they are calculated.

Figure 4-10   Sample series *RL* circuit problem: (a) circuit; (b) procedure for finding circuit values; (c) circuit triangles.

triangles used to solve series *RC* circuits problems and Figure 4-12 presents an example of a series *RC* circuit problem.

Series *RLC* circuits.   Series *RLC* circuits such as the one shown in Figure 4-13 have resistance (R), inductance (L), and capacitance (C). The total reactance (X_T) is found by subtracting the smaller reactance (X_L or X_C) from the larger one. Reactive

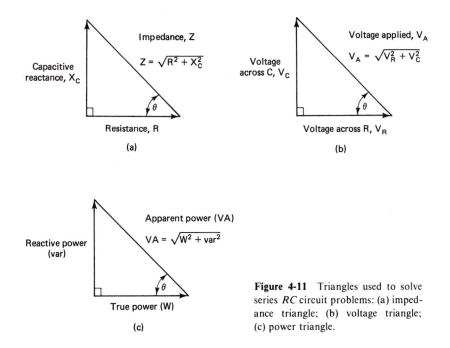

Figure 4-11 Triangles used to solve series *RC* circuit problems: (a) impedance triangle; (b) voltage triangle; (c) power triangle.

voltage (V_X) is found by obtaining the difference between V_L and V_C. The effects of the capacitance and inductance are 180° out of phase with each other. Right triangles, as shown in Figure 4-14, are used to show the simplified relationships of the circuit values.

Parallel ac circuits.   Parallel ac circuits are analyzed using procedures similar to those for series ac circuits. The methods used with parallel ac circuits are somewhat different from those of series ac circuits. For instance, impedance (Z) of a parallel circuit is *less than* any individual branch values of resistance, inductive reactance, or capacitive reactance. There is no impedance triangle for parallel circuits since Z is smaller than R, X_L, or X_C. A right triangle may be drawn to show the currents in the branches of parallel circuit.

The voltage of a parallel ac circuit is the same across each branch. The currents through the branches of a parallel ac circuit are shown by a right triangle called a *current triangle* (see Figure 4-15). The current through the capacitor (I_C) leads the current through the resistor (I_R) by 90°. The current through the inductor (I_L) lags I_R by 90°. Therefore, I_L and I_C are 180° out of phase and are subtracted to find the total reactive current (I_X). Since these values form a right triangle, the total current of parallel ac circuits may be found by using the formula

$$I_T = \sqrt{I_R^2 + I_X^2}$$

For components connected in parallel, an impedance triangle cannot be used. The method used to find impedance is an *admittance triangle* (see Figure 4-15). The following quantities are plotted on the triangle: admittance—$Y = 1/Z$, conductance—

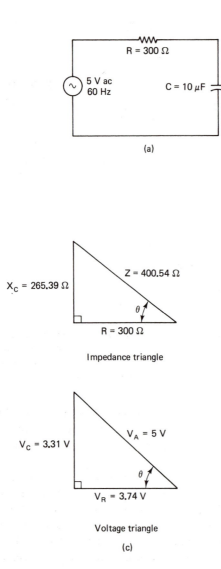

(a)

Impedance triangle

$X_C = 265.39 \ \Omega$

$Z = 400.54 \ \Omega$

θ

$R = 300 \ \Omega$

Voltage triangle

$V_C = 3.31 \ V$

$V_A = 5 \ V$

θ

$V_R = 3.74 \ V$

(c)

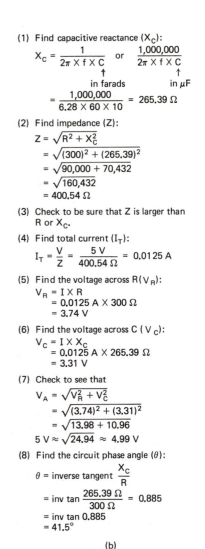

(1) Find capacitive reactance (X_C):
$$X_C = \frac{1}{2\pi \times f \times C} \quad \text{or} \quad \frac{1,000,000}{2\pi \times f \times C}$$
$$\uparrow \qquad\qquad\qquad \uparrow$$
$$\text{in farads} \qquad\quad \text{in } \mu F$$
$$= \frac{1,000,000}{6.28 \times 60 \times 10} = 265.39 \ \Omega$$

(2) Find impedance (Z):
$$Z = \sqrt{R^2 + X_C^2}$$
$$= \sqrt{(300)^2 + (265.39)^2}$$
$$= \sqrt{90,000 + 70,432}$$
$$= \sqrt{160,432}$$
$$= 400.54 \ \Omega$$

(3) Check to be sure that Z is larger than R or X_C.

(4) Find total current (I_T):
$$I_T = \frac{V}{Z} = \frac{5 \ V}{400.54 \ \Omega} = 0.0125 \ A$$

(5) Find the voltage across R(V_R):
$$V_R = I \times R$$
$$= 0.0125 \ A \times 300 \ \Omega$$
$$= 3.74 \ V$$

(6) Find the voltage across C (V_C):
$$V_C = I \times X_C$$
$$= 0.0125 \ A \times 265.39 \ \Omega$$
$$= 3.31 \ V$$

(7) Check to see that
$$V_A = \sqrt{V_R^2 + V_C^2}$$
$$= \sqrt{(3.74)^2 + (3.31)^2}$$
$$= \sqrt{13.98 + 10.96}$$
$$5 \ V \approx \sqrt{24.94} \approx 4.99 \ V$$

(8) Find the circuit phase angle (θ):
$$\theta = \text{inverse tangent } \frac{X_C}{R}$$
$$= \text{inv tan } \frac{265.39 \ \Omega}{300 \ \Omega} = 0.885$$
$$= \text{inv tan } 0.885$$
$$= 41.5°$$

(b)

Figure 4-12 Sample series *RC* circuit problem: (a) circuit; (b) procedure for finding circuit values; (c) circuit triangles.

Figure 4-13 Series *RLC* circuit.

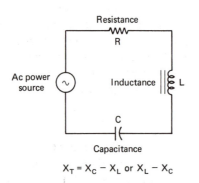

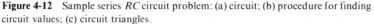

$$X_T = X_C - X_L \text{ or } X_L - X_C$$

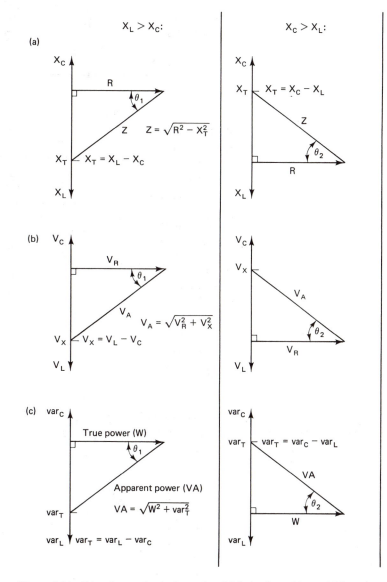

Figure 4-14 Triangles used to solve series *RLC* circuit problems: (a) impedance triangles; (b) voltage triangles; (c) power triangles.

$G = 1/R$, inductive susceptance—$B_L = 1/X_L$, and capacitive susceptance—$B_C = 1/X_C$. Notice that these quantities are the inverse of each type of opposition to ac current. Since total impedance (Z) is the smallest quantity in a parallel ac circuit, its reciprocal ($1/Z$) becomes the largest quantity on the admittance triangle. The values on the triangle are in units of *seimen* or *mho* ("ohm" spelled backward). Parallel ac circuit examples are shown in Figures 4-15 and 4-16.

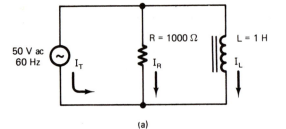

(a)

Finding circuit values:

1. Find inductive reactance (X_L):

$$X_L = 2\pi \cdot f \cdot L$$
$$= 6.28 \times 60 \times 1$$
$$= 376.8\ \Omega$$

2. Find current through $R\ (I_R)$:

$$I_R = \frac{E}{R} = \frac{50\ V}{1000\ \Omega} = 0.05\ A$$

3. Find current through $L\ (I_L)$:

$$I_L = \frac{E}{X_L} = \frac{50\ V}{376.8\ \Omega} = 0.133\ A$$

4. Find total current (I_T):

$$I_T = \sqrt{I_R^2 + I_L^2}$$
$$= \sqrt{0.05^2 + 0.133^2}$$
$$= \sqrt{0.0025 + 0.0177}$$
$$= \sqrt{0.0202}$$
$$= 0.142\ A$$

5. Check to see that I_T is larger than I_R or I_L.

6. Find impedance (Z):

$$Z = \frac{V}{I_T} = \frac{50\ V}{0.142\ A} = 352.1\ \Omega$$

7. Check to see that Z is less than R or X_L.

8. Find circuit phase angle (θ):

$$\sin\theta = \frac{\text{opposite}}{\text{hypotenuse}} = \frac{I_L}{I_T} = \frac{0.133}{0.142}$$
$$= 0.937$$
$$\theta = \text{inverse sine } 0.937$$
$$= 70°$$

(b)

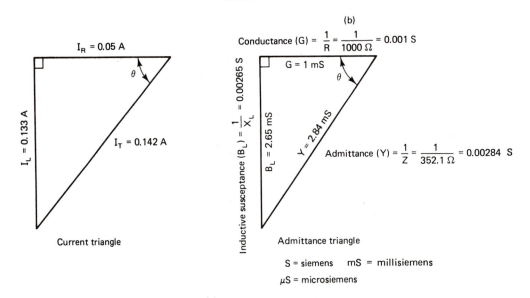

Conductance $(G) = \dfrac{1}{R} = \dfrac{1}{1000\ \Omega} = 0.001\ S$

$G = 1\ mS$

Admittance $(Y) = \dfrac{1}{Z} = \dfrac{1}{352.1\ \Omega} = 0.00284\ S$

$I_R = 0.05\ A$

$I_L = 0.133\ A$

$I_T = 0.142\ A$

Current triangle

Inductive susceptance $(B_L) = \dfrac{1}{X_L} = 0.00265\ S$

$B_L = 2.65\ mS$

$Y = 2.84\ mS$

Admittance triangle

S = siemens   mS = millisiemens

μS = microsiemens

(c)

Figure 4-15 Sample parallel *RL* circuit problem: (a) circuit; (b) procedure for finding circuit values; (c) circuit triangles.

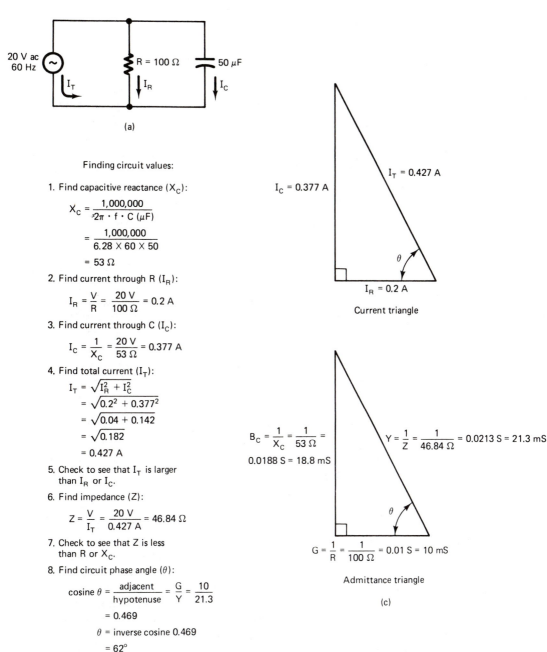

(a)

Finding circuit values:

1. Find capacitive reactance (X_C):

$$X_C = \frac{1,000,000}{2\pi \cdot f \cdot C\ (\mu F)}$$

$$= \frac{1,000,000}{6.28 \times 60 \times 50}$$

$$= 53\ \Omega$$

2. Find current through R (I_R):

$$I_R = \frac{V}{R} = \frac{20\ V}{100\ \Omega} = 0.2\ A$$

3. Find current through C (I_C):

$$I_C = \frac{1}{X_C} = \frac{20\ V}{53\ \Omega} = 0.377\ A$$

4. Find total current (I_T):

$$I_T = \sqrt{I_R^2 + I_C^2}$$

$$= \sqrt{0.2^2 + 0.377^2}$$

$$= \sqrt{0.04 + 0.142}$$

$$= \sqrt{0.182}$$

$$= 0.427\ A$$

5. Check to see that I_T is larger than I_R or I_C.

6. Find impedance (Z):

$$Z = \frac{V}{I_T} = \frac{20\ V}{0.427\ A} = 46.84\ \Omega$$

7. Check to see that Z is less than R or X_C.

8. Find circuit phase angle (θ):

$$\text{cosine } \theta = \frac{\text{adjacent}}{\text{hypotenuse}} = \frac{G}{Y} = \frac{10}{21.3}$$

$$= 0.469$$

$$\theta = \text{inverse cosine } 0.469$$

$$= 62°$$

(b)

$I_C = 0.377\ A$ $I_T = 0.427\ A$

θ

$I_R = 0.2\ A$

Current triangle

$$B_C = \frac{1}{X_C} = \frac{1}{53\ \Omega} =$$

$0.0188\ S = 18.8\ mS$

$$Y = \frac{1}{Z} = \frac{1}{46.84\ \Omega} = 0.0213\ S = 21.3\ mS$$

θ

$$G = \frac{1}{R} = \frac{1}{100\ \Omega} = 0.01\ S = 10\ mS$$

Admittance triangle

(c)

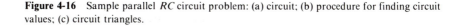

Figure 4-16 Sample parallel *RC* circuit problem: (a) circuit; (b) procedure for finding circuit values; (c) circuit triangles.

POWER IN AC CIRCUITS

Power quantities of ac circuits are found by using *power triangles*. Power triangles are the same for series and parallel ac circuits [see Figures 4-9(c), 4-11(c), and 4-14(c)]. The power *delivered* to an ac circuit is called *apparent power* and is equal to voltage multiplied by current ($V \times I$). The unit of measurement of apparent power is the voltampere (VA) or kilovoltampere (kVA). A voltmeter and ammeter may be used to measure apparent power in ac circuits. The values of voltage and current are multiplied to determine the apparent power of a circuit. The actual power converted to another form of energy by the load is measured with a wattmeter and is called *true power*. The ratio of true power converted in a circuit to apparent power delivered to an ac circuit is called the *power factor*. The power factor is found by using the formula

$$PF = \frac{\text{true power (W)}}{\text{apparent power (VA)}} \quad \text{or} \quad \% \text{ PF} = \frac{W}{VA} \times 100$$

The term W is the true power in watts and VA is the apparent power in voltamperes. The highest power factor of an ac circuit is 1.0 or 100% and is called *unity* power factor. Unity power factor is a characteristic of ac circuits with resistance only.

The phase angle (θ) of an ac circuit determines the power factor. Phase angle is the angular separation between voltage applied to an ac circuit and total current flow through the circuit. Increased inductive reactance or capacitive reactance causes a larger phase angle. In purely inductive or capacitive circuits, a 90° phase angle causes a power factor of *zero*. Power factor varies according to the relative values of resistance and reactance in an ac circuit.

There are two types of power that affect power conversion in ac circuits. Power conversion in the resistive part of the circuit is called *active power* or true power. True power is measured (in watts) with a wattmeter. The other type of power is present when a circuit has inductance or capacitance. It is called *reactive power* and is 90° out of phase with true power. Reactive power is measured in voltamperes-reactive or var and is sometimes considered "unused" power in ac circuits.

The power triangles shown in Figures 4-9(c), 4-11(c), and 4-14(c) have true power (watts) shown as horizontal lines. Reactive power (var) is plotted at a 90° angle from true power. Voltamperes or apparent power (VA) form the hypotenuse of the power triangles. The formulas that are used for calculating values of a power triangle are shown in Figure 4-17. Power triangles are similar to impedance triangles and the voltage triangles for series ac circuits and current triangles and admittance triangles for parallel ac circuits. Circuit phase angle may be found by using power triangles also. Each type of circuit triangle has a horizontal line used to show the *resistive* part of the circuit. The vertical lines are used to show the *reactive* part of the circuit. The hypotenuse of each right triangle is used to show *total* values of the circuit. The length of the hypotenuse of a triangle depends on the relative amounts of resistance and reactance in a circuit. Circuit triangles are extremely helpful for analyzing ac circuits. An understanding of basic trigonometry and right triangles is necessary in analyzing ac circuits. Refer to Appendix 2 for a review of trigonometry and right triangles.

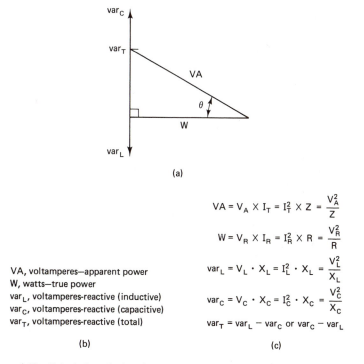

VA, voltamperes—apparent power
W, watts—true power
var$_L$, voltamperes-reactive (inductive)
var$_C$, voltamperes-reactive (capacitive)
var$_T$, voltamperes-reactive (total)

(b)

$$VA = V_A \times I_T = I_T^2 \times Z = \frac{V_A^2}{Z}$$

$$W = V_R \times I_R = I_R^2 \times R = \frac{V_R^2}{R}$$

$$var_L = V_L \cdot X_L = I_L^2 \cdot X_L = \frac{V_L^2}{X_L}$$

$$var_C = V_C \cdot X_C = I_C^2 \cdot X_C = \frac{V_C^2}{X_C}$$

$$var_T = var_L - var_C \text{ or } var_C - var_L$$

(c)

Figure 4-17 Calculation of values for a power triangle: (a) power triangle for an *RLC* circuit; (b) abbreviations used; (c) formulas for calculating true power, apparent power, and reactive power.

BASIC AC GENERATOR OPERATION

When conductors move through a magnetic field or a magnetic field is moved past conductors, an induced current is developed. The current that is induced into the conductors produces an induced electromotive force or voltage. This is the basic operational principle of all mechanical generators.

Left-hand rule of current flow.   The direction of current flow through a moving conductor within a magnetic field can be determined by using the *left-hand rule*. Refer to Figure 4-18 and use the left hand in the following manner:

1. Arrange the thumb, forefinger, and middle finger so that they are at approximately right angles to one another.
2. Point the thumb in the direction of conductor movement.
3. Point the forefinger in the direction of the magnetic lines of force (from north to south).
4. The middle finger will now point in the direction of induced electron current flow (negative to positive).

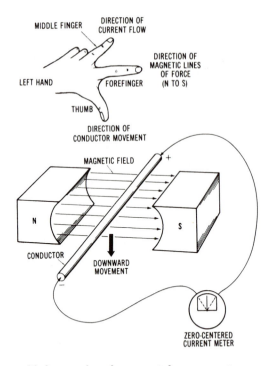

Figure 4-18 Left-hand rule of current flow.

Voltage development in generators.   When a conductor moves across magnetic lines of force, an induced current is developed through the conductor. This causes a difference in potential or voltage across the ends of the conductor. The induced voltage value may be changed by modifying either the strength of the magnetic field or the relative speed of conductor movement through the magnetic field. If the magnetic field is made stronger, more voltage is induced. If the conductor is moved at a faster speed, more voltage is induced. Similarly, if more conductors are concentrated within the magnetic field, a greater voltage is developed. These rules of electromagnetic induction, which were discussed in Chapter 1, are very important for the operation of mechanical generators.

SINGLE-PHASE AC GENERATORS

Single-phase electrical power is produced by mechanical generators which are commonly called alternators. The principle of operation of a single-phase alternator relies on electromagnetic induction. In order for a generator to convert mechanical energy into electrical energy, three conditions must exist:

1. There must be a magnetic field.
2. There must be a group of conductors adjacent to the magnetic field.
3. There must be relative motion between the magnetic field and the conductors.

These conditions are necessary in order for electromagnetic induction to occur.

Generator construction. Generators used to produce electrical energy require some form of mechanical energy input. This mechanical energy is used to produce relative motion between the electrical conductors and the magnetic field of the generator. Figure 4-19 show the basic parts and construction of an ac generator. Mechanical generators have a stationary part and a rotating part. The stationary part is called the *stator* and the rotating part is called the *rotor*. A generator has magnetic *field poles* which produce north and south polarities. Also, a generator must have a method of producing mechanical energy in the form of rotary motion. A *prime mover* used for this purpose is connected to the rotor shaft of a generator. There must be a method of electrically connecting the conductors of the rotor to the external circuit. This is done by a *slip-ring/brush assembly* with stationary brushes made of carbon and graphite. The slip rings used on ac generators are ordinarily made of copper and are permanently mounted on the generator shaft. The two slip rings connect to the ends of the *conductor loops* or rotor windings. When an external load is connected to the generator, a closed circuit is completed. With all of these generator parts functioning together, electromagnetic induction can take place and electrical energy can be produced.

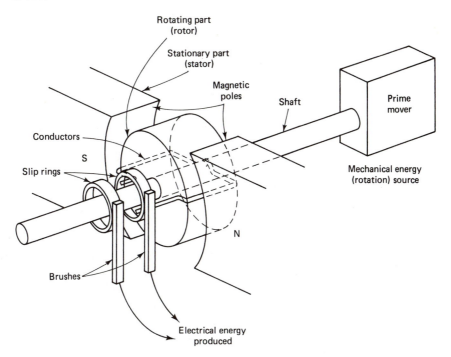

Figure 4-19 Basic generator construction. A generator must have (1) a magnetic field, (2) conductors, and (3) a source of mechanical energy (rotation).

Generating ac voltage. Figure 4-20 shows a magnetic field developed by permanent magnets used to illustrate the operational principle of a mechanical generator. Conductors that can be rotated on a center axis are placed within the

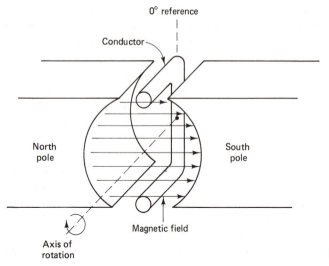

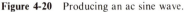

Figure 4-20   Producing an ac sine wave.

magnetic field. They are connected to the external load by means of a slip-ring/brush assembly. The 0° reference position shown on the diagram is used as a starting point for the following discussion of ac voltage generation.

Figure 4-21 illustrates the basic operation of an ac generator. In position A [Figure 4-21(a)], the conductors are positioned so that the minimum amount of magnetic lines of force are "cut" by the conductors as they rotate. This is the 0° reference position. Minimum (zero) current is induced into the conductors at position A. Observe position A of the resulting output waveform [Figure 4-21(f)]. The resulting current flow through the load at position A is zero.

If the conductors are rotated 90° in a clockwise direction to position B [Figure 4-21(b)], they pass from the minimum number of magnetic lines of force to the most concentrated area of the magnetic field. At position B, the induced current is maximum, as shown by the resulting output waveform diagram of Figure 4-21(f). Note that the induced current rises gradually from the zero reference line to a maximum value at position B.

As the conductors are rotated another 90° to position C [Figure 4-21(c)], the induced current becomes zero again. No current flows through the load at this position. Note in the diagram of the resulting output of Figure 4-21(f) how the induced current drops gradually from maximum to zero. This part of the induced current (from 0 to 180°) is called the *positive alternation*. Each value of the induced current, as the conductors rotate from the 0° position to the 180° position, is in a positive direction. This action could be observed visually if a meter were connected in place of the load.

When the conductors are rotated another 90° to position D [Figure 4-21(d)], they pass once again through the most concentrated portion of the magnetic field. Maximum current is induced into the conductors at this position. However, the direction of the induced current is in the opposite direction compared with position B. At the 270° position, the induced current is maximum in a negative direction.

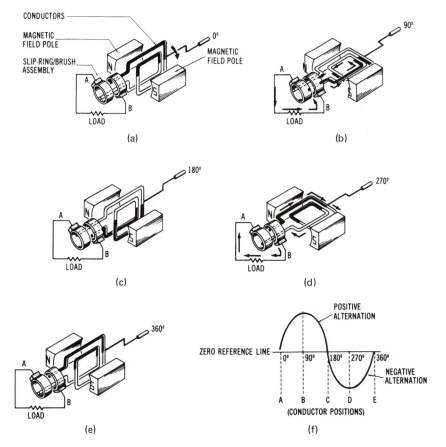

Figure 4-21 Basic parts and operational principle of a mechanical generator: (a) position A; (b) position B; (c) position C; (d) position D; (e) position E; (f) resulting output waveform.

Observe closely at this point the direction of induced current through the conductors, the slip rings, and the external load. As the conductors are rotated to position E (same as at position A), the induced current is minimum once again. Note in the diagram of Figure 4-21(e) how the induced current decreases from its maximum negative value back to zero again (at the 360° position). The induced current waveform from 180 to 360° is called the *negative alternation.* The complete output, which shows the induced current through the load, is called a single-phase alternating-current waveform. As the conductors continue to rotate through the magnetic field, the cycle is repeated.

AC sine-wave values.   The induced current developed by the method discussed above produces a sinusoidal waveform or *sine wave.* This waveform is referred to as a sine wave due to its mathematical origin, based on the trigonometric

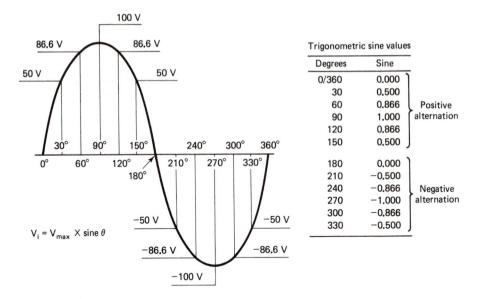

| Trigonometric sine values | | |
Degrees	Sine	
0/360	0.000	Positive alternation
30	0.500	
60	0.866	
90	1.000	
120	0.866	
150	0.500	
180	0.000	Negative alternation
210	−0.500	
240	−0.866	
270	−1.000	
300	−0.866	
330	−0.500	

Figure 4-22   Ac sine-wave values.

sine function. The current induced into the conductors, shown in Figure 4-22, varies as the sine of the angle of rotation between the conductors and the magnetic field. This induced current develops a voltage or potential difference across the ends of the conductor loops. The instantaneous voltage induced into a single conductor can be expressed as

$$V_i = V_{max} \times \sin \theta$$

where V_i = instantaneous induced voltage
V_{max} = maximum induced voltage
θ = angle of conductor rotation from the zero reference

For example, at the 30° position (Figure 4-22), if the maximum voltage is 100 V, then $V_i = 100 \times \sin \theta$. Refer to Appendix 2 to find that the sine of 30° = 0.5. Then

$$V_i = 100 \times 0.5$$

$$= 50 \text{ V}$$

Observe in Appendix 2 that sine values are positive in quadrant 1 (0–90°) and quadrant 2 (90–180°) and negative in quadrant 3 (180–270°) and quadrant 4 (270–360°). Thus, at the 240° position,

$$V_i = 100 \times \sin 240°$$

$$= 100 \times -0.866$$

$$= -86.6 \text{ V}$$

Types of Single-Phase AC Generators

Although much single-phase electrical energy is used, particularly in the home, very little is produced by single-phase alternators. The single-phase electrical power used in the home is usually developed by three-phase alternators and then converted to single-phase ac by the power distribution system. There are two basic methods that can be used to produce single-phase alternating current. One method is called the rotating-armature method and the other is the rotating-field method. Keep in mind that the term "alternator" refers to any type of ac generator.

Rotating-armature method.   In the rotating-armature method, shown in Figure 4-23, an alternating-current voltage is induced into the conductors of the rotating part of the machine. The electromagnetic field is developed by a set of stationary pole pieces. Relative motion between the conductors and the magnetic field is provided by a prime mover or mechanical-energy source connected to the rotor shaft. Prime movers may be steam turbines, gas turbines, hydraulic turbines, gasoline engines, diesel engines, or possibly electric motors. All generators convert mechanical energy into electrical energy, as shown in Figure 4-24. Only small power ratings can be used with the rotating-armature type of alternator. The major disadvantage of this

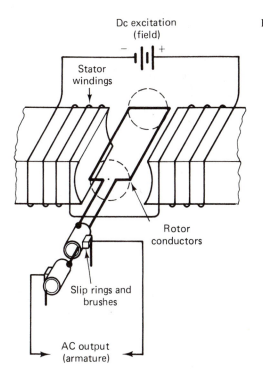

Figure 4–23   Rotating-armature method.

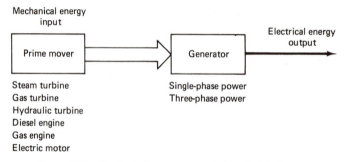

Mechanical energy
input

Prime mover → Generator

Electrical energy
output

Steam turbine
Gas turbine
Hydraulic turbine
Diesel engine
Gas engine
Electric motor

Single-phase power
Three-phase power

Figure 4-24   Mechanical energy converted to electrical energy.

method is that the ac voltage is extracted from a slip-ring/brush assembly (see Figure 4-23). A high voltage could produce tremendous sparking or arc-over between the brushes and the slip rings. The maintenance involved in replacing brushes and repairing the slip-ring commutator assembly would be very time consuming and expensive. Therefore, this method is used primarily for alternators with lower power ratings.

Rotating-field method.   The rotating-field method, shown in Figure 4-25, is used for alternators capable of producing larger amounts of power. The dc excitation voltage that develops the magnetic field is applied to the rotating portion of the

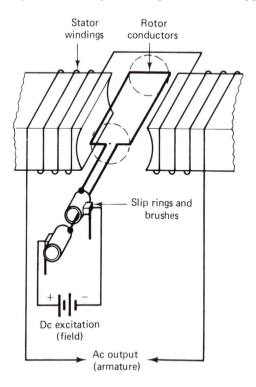

Figure 4-25   Rotating-field method.

Stator
windings

Rotor
conductors

Slip rings and
brushes

Dc excitation
(field)

Ac output
(armature)

machine. The ac voltage is induced into the stationary conductors of the machine. Since the dc excitation voltage is a much lower value than the ac voltage that is produced, maintenance problems associated with the slip-ring brush assembly are minimized. In addition, the conductors of the stationary portion of the machine may be larger so as to handle more current since they do not rotate.

ELECTROMAGNETIC INDUCTION IN AC GENERATORS

Voltage is developed in electrical generators due to relative motion between a magnetic field and conductors. The voltage developed is called an "induced" voltage because it occurs without actual physical contact between the magnetic field and conductors. A general statement of Faraday's law of electromagnetic induction is: "The magnitude of voltage induced into a single turn of wire is proportional to the rate of change of the lines of force passing through that turn." Thus the amount of induced voltage may be increased by increasing the magnetic field strength (number of lines of force) or by increasing the speed or relative motion between the magnetic field and conductor.

The instantaneous voltage (V_i) induced into a conductor may be expressed in terms of an average magnetic flux density and the relative speed (velocity) between the magnetic field and conductor. For a specific length of conductor, the instantaneous induced voltage (V_i) is expressed as

$$V_i = \frac{1}{5} BLV \times 10^{-8}$$

where B = magnetic flux density in lines/in^2
L = length of conductor adjacent to the magnetic flux in inches
V = velocity (speed) in ft/min between the conductor and field
10^{-8} = a constant equal to the number of flux lines a single conductor must link each second to induce a voltage of 1 V

For example, assume that a conductor is 24 in. long and removed through a uniform magnetic field of 30,000 lines/in^2. The distance traveled in one second is 600 in. The instantaneous voltage may be found as follows:

$$V_i = \frac{1}{5} BLV \times 10^{-8}$$

$$= \frac{1}{5} \times (30,000 \text{ lines/in}^2) \times (24 \text{ in.}) \times (600 \text{ in./s} \times 60 \text{ s/min} \times 1 \text{ ft/12 in.}) \times 10^{-8}$$

$$= (0.2)(30,000)(24)(600 \times 60 \times 0.0833)(10^{-8})$$

$$= 4.32 \text{ V}$$

The preceding example assumes a uniform flux density of the magnetic field and a uniform relative motion between the magnetic field and conductor.

Lenz's law in ac generation.   The movement of a conductor through a magnetic field is the result of mechanical energy. The electrical energy that is produced by electromagnetic induction requires a mechanical energy input in accordance with the *law of conservation of energy*. This states that "energy can neither be created nor destroyed." Lenz's law relates to the energy conversion which takes place during electromagnetic induction in generators. The general statement of Lenz's law is: "In all cases of electromagnetic induction, the induced voltage causes a current flow in such a direction that its magnetic field opposes the change that produces it." Thus an induced voltage produces a current flow in a direction that opposes the change (in the form of motion) that produces the current flow.

Calculation of average induced voltage.   The average induced voltage that is produced by an ac generator may be calculated using the procedure described below. The total number of turns (N) must be determined as follows:

$$N = CN_c$$

where C = total number of armature coils
N_c = number of turns of wire per coil

A typical armature could have 100 armature coils (C) and two turns per coil (N_c). Therefore, the total number of turns of wire on the armature would be

$$N = C \times N_c$$
$$= 100 \times 2$$

Keep in mind that the armature of an ac generator is defined as the part of the machine into which current is induced. The armature can be either the rotor or stator portion of the machine, depending on the type of construction employed.

The average induced voltage (V_{avg}) for an ac generator may be found by applying the equation

$$V_{avg} = 4\emptyset Nf \times 10^{-8} \text{ V}$$

where $\emptyset$ = number of magnetic flux lines per pole
N = total number of armature turns
f = frequency in hertz
10^{-8} = a constant

Also, since effective or root-mean-square (rms) voltage (V_{eff}) = $V_{avg} \times 1.11$, the generated rms voltage of an ac generator may be determined as follows:

$$V_{eff} = 4.44\emptyset Nf \times 10^{-8}$$

For example, assume that an armature has 5×10^4 magnetic flux lines per pole, 200 armature turns, and operates at a frequency of 60 Hz. The effective voltage generated would be

$$V_{eff} = 4.44\emptyset Nf \times 10^{-8}$$

$$= (4.44)(5 \times 10^4)(200)(60)(10^{-8})$$

$$= 266.4 \text{ V}$$

For three-phase ac machines, the number of turns per phase (N_p) must be considered as

$$N_p = \frac{C \times N_c}{P}$$

where P is the number of phases. The N_p value is then substituted for N in the equation above.

Assume that a three-phase generator has 6×10^4 magnetic flux lines per pole, 20 armature turns with two turns per coil, and separates at a frequency of 60 Hz. The effective generated voltage is found as follows:

$$N_p = \frac{C \times N_c}{P} = \frac{120 \times 2}{3} = \frac{240}{3} = 80 \text{ turns/phase}$$

and

$$V_{\text{eff}} = 4.44\emptyset N_p f \times 10^{-8}$$

$$= (4.44)(6 \times 10^4)(80)(60)(10^{-8})$$

$$= 127.8 \text{ V}$$

THREE-PHASE AC GENERATORS

The vast majority of electrical power produced in the United States is three-phase power. A steam turbine generator used at a power plant is shown in Figure 4-26. Most generators that produce three-phase power look similar to this one. Due to their large power ratings, three-phase generators utilize the rotating field method. A typical

Figure 4-26   *Steam turbine generator used at a power plant. (Courtesy of General Electric Co.)*

TURBINE-GENERATOR				
Steam Turbine				

Rating:	66,000 kW		3600 rpm	21 stages
Steam: Pressure	1250 psig		Temp. 950°F	Exhaust pressure 1.5 in. Hg abs.

Generator					
Hydrogen cooled			Rating	Capability	Capability
Type ATB	2 poles	60 cycles	Gas pressure	30 psig	15 psig	0.5 psig
3 ph. Y Connected for		13,800 V	kVA	88,235	81,176	70,588
Excitation		250 V	Kilowatts	75,000	69,000	60,000
Temp. rise guaranteed not to exceed			Armature amp.	3691	3396	2953
45°C on armature by detector			Field amp.	721	683	626
74°C on field by resistance			Power factor	0.85	0.85	0.85

Figure 4-27   Nameplate data for a three-phase turbine generator.

three-phase generator in a power plant might have 250 to 500 V dc excitation applied to the rotating field through the slip-ring/brush assembly. The typical output of 13.8 kV ac is induced into the stationary conductors of the machine.

Commercial electrical power systems use many three-phase alternators connected in parallel to supply their regional load requirements. Normally, industrial loads represent the largest portion of the load on our power systems. The residential (home) and commercial loads are somewhat less. Due to the vast load that has to be met by electrical power systems, three-phase generators have high power ratings. Nameplate data for a typical commercial three-phase turbine-generator are shown in Figure 4-27. Note the large values for each of the nameplate ratings.

Generation of Three-Phase Voltage

The basic construction of a three-phase ac generator is shown in Figure 4-28, with its resulting voltage output waveforms given in Figure 4-29. Note that three-phase generators must have at least six stator poles or two poles for each phase. The three-phase generator shown in the drawing is a rotating-field type of generator. The magnetic field is developed electromagnetically by a direct-current voltage source. The dc voltage is applied from an external power source through a slip-ring/brush assembly to the windings of the rotor. The magnetic polarities of the rotor as shown are north at the top and south at the bottom of the illustration. The magnetic lines of force would be developed around the outside of the electromagnetic rotor assembly.

Through the process of electromagnetic induction, a current can be induced into each of the stationary stator coils of the generator. Since the beginning of phase A is physically located 120° from the beginning of phase B, the induced currents will be

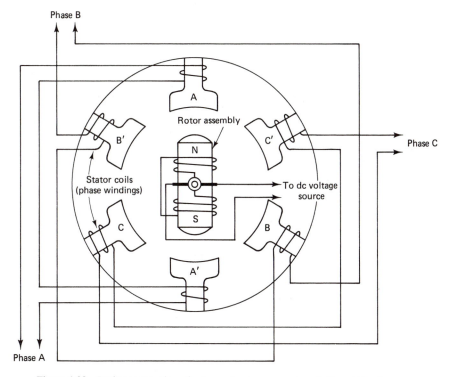

Figure 4-28 Basic construction of a three-phase ac generator. A, B, and C poles are 120° apart; A′, B′, and C′ poles are 120° apart; A′ and A, B′ and B, and C′ and C are 180° apart.

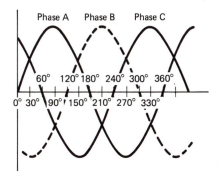

Figure 4-29 Output waveforms of a three-phase ac generator. Note that there is a 120° phase separation.

120° apart. Similarly, the beginning of phase B and the beginning of phase C are located 120° apart. Thus the voltages developed due to electromagnetic induction are 120° apart, as shown in Figure 4-30. Voltages are developed in each stator winding as the electromagnetic field rotates within the enclosure that houses the stator coils. A prime mover provides the mechanical energy to produce rotation of the rotor assembly.

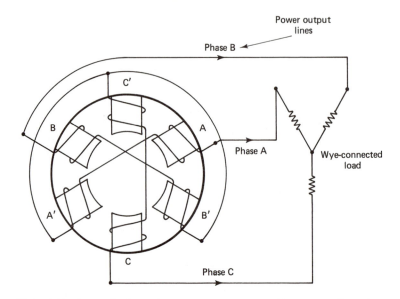

Figure 4-30   Stator of a three-phase ac generator connected in a wye configuration.

Three-Phase Connection Methods

In Figure 4-28, poles A', B', and C' represent the beginnings of each of the phase windings of the alternator. Poles A, B, and C represent the ends of each of the phase windings. There are two methods that may be used to connect these windings together. These methods are called wye and delta connections. They are basic to all three-phase systems.

Three-phase wye-connected generators.   The windings of a three-phase ac generator can be connected in a wye configuration by connecting the beginnings *or* the ends of the windings together to make a common connection. The other ends of the windings become the three-phase power output lines from the generator. A three-phase wye-connected generator is illustrated in Figure 4-30. Notice that the beginnings of the windings (poles A', B', and C') are connected together. The ends of the windings (poles A, B, and C) are the three-phase power output lines that are connected to the load to which the generator supplies power. The load is represented by resistances connected in a wye configuration.

Three-phase delta-connected generators.   The windings of a three-phase ac generator may also be connected in a delta arrangement, as shown in Figure 4-31. In the delta configuration, the beginning of one phase winding is connected to the end of the adjacent phase winding. Thus the beginnings and ends of all adjacent phase windings are connected together. The three power output lines are brought out

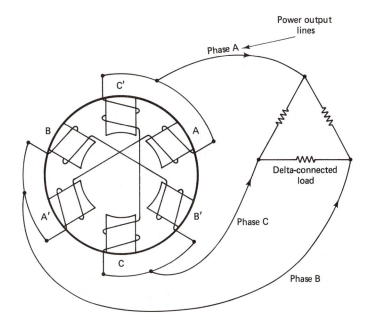

Figure 4-31 Stator of a three-phase ac generator connected in a delta configuration.

of the ac generator from the connections of the beginning–end winding combinations. The load illustrated in Figure 4-31 represents resistances connected in a delta configuration.

Advantages of Three-Phase Power

Three-phase power is used primarily for industrial and commercial applications. Many types of industrial equipment use three-phase alternating-current power because the power produced by a three-phase ac voltage source is less pulsating than single-phase ac power. This effect is noted by observing that a peak voltage occurs every 120° in the three-phase waveforms of Figure 4-29. A single-phase voltage source has a peak voltage output only once every 360°. This comparison is somewhat similar to comparing the power developed by an eight-cylinder automobile engine to the power developed by a four-cylinder engine. The eight-cylinder engine provides smoother, less pulsating power. The effect of smoother power development on the operation of electric motors (with three-phase voltage applied) is that it produces a more uniform torque in the motor. This factor is particularly important for large motors that are used in industry.

Three separate single-phase voltages can be derived from a three-phase power transmission line. It is also more economical to distribute three-phase ac power from power plants to consumers that are located a considerable distance away. Fewer conductors are required to distribute three-phase voltage than three separate single-phase voltages. Also, equipment that uses three-phase power is physically smaller in size than comparable single-phase equipment.

Voltage and Current Relationships of Three-Phase AC Generators

In the schematic of a three-phase wye connection of Figure 4-32, the beginnings or the ends of each winding are connected together. The other sides of the windings become the ac power output lines from the generator. The voltage across the ac power output lines (V_L) is equal to the square root of 3 (1.73) multiplied by the voltage across the phase windings (V_P), or

$$V_L = V_P \times 1.73$$

The line currents (I_L) are equal to the phase currents (I_P), or

$$I_L = I_P$$

In the schematic of a three-phase delta connection of Figure 4-33, the end of one phase winding is connected to the beginning of the adjacent phase winding. The line voltages (V_L) are equal to the phase voltage (V_P). The line currents (I_L) are equal to the phase currents (I_P) multipled by 1.73.

Figure 4-32   Schematic of a three-phase wye connection.

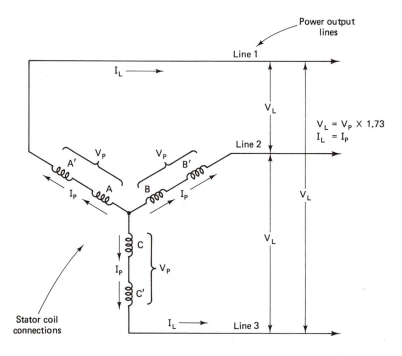

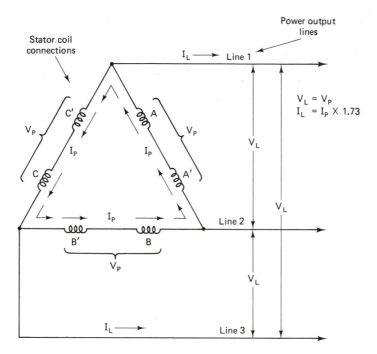

Figure 4-33 Schematic of a three-phase delta connection.

Power Relationships of Three-Phase AC Generators

The power developed in each phase (P_P) for either a wye or a delta system is expressed as

$$P_P = V_P \times I_P \times \text{PF}$$

where PF is the power factor (phase angle between voltage and current) of the load. Other three phase quantities are listed below:

Voltamperes per phase: $\text{VA}_P = V_P I_P$

Total voltamperes: $\text{VA}_T = 3V_P I_P$
$$= 1.73 V_L I_L$$

Power factor: $\text{PF} = \dfrac{\text{true power (W)}}{1.73 V_L I_L}$
$$= \dfrac{\text{W}}{3V_P I_P}$$

Total power: $P_T = 3V_P I_P \times \text{PF}$
$$= 1.73 V_L I_L \times \text{PF}$$

Three-Phase AC Generator Sample Problems

Assume that a three-phase ac generator is connected first in a wye configuration and then in a delta configuration. Its given values are as follows: phase voltage = 120 V, phase current = 20 A, frequency = 60 Hz, power output = 6 kW.

With the stator coils connected in a wye configuration, the following values are obtained:

Line voltage: $V_L = V_P \times 1.73$
$$= 120 \times 1.73$$
$$= 208 \text{ V}$$

Line current: $I_L = I_P$
$$= 20 \text{ A}$$

Voltamperes per phase: $\text{VA}_P = V_P \times I_P$
$$= 120 \times 20$$
$$= 2400 \text{ VA}$$

Total voltamperes: $\text{VA}_T = 3 \times V_P \times I_P$
$$= 3 \times 120 \times 20$$
$$= 7200 \text{ VA}$$

Power factor: $\text{PF} = \dfrac{\text{W}}{\text{VA}_T} = \dfrac{6000}{7200} = 0.83 \quad \text{or} \quad 83\%$

Power per phase: $P_P = V_P \times I_P \times \text{PF}$
$$= 120 \times 20 \times 0.83$$
$$= 1992 \text{ W}$$

Total power: $P_T = 3 \times V_P \times I_P \times \text{PF}$
$$= 3 \times 120 \times 20 \times 0.83$$
$$= 5976 \text{ W}$$

With the stator coils of the same three-phase ac generator connected in a delta configuration, these values are obtained:

Line voltage: $V_L = V_P$
$$= 120 \text{ V}$$

Line current: $I_L = I_P \times 1.73$
$$= 20 \text{ A} \times 1.73$$
$$= 34.6 \text{ A}$$

Voltamperes per phase: $\text{VA}_P = V_P \times I_P$
$$= 120 \text{ V} \times 20 \text{ A}$$
$$= 2400 \text{ VA}$$

Total voltamperes: $\text{VA}_T = 3 \times V_P \times I_P$
$$= 3 \times 120 \text{ V} \times 20 \text{ A}$$
$$= 7200 \text{ VA}$$

Power factor: $\text{PF} = \dfrac{\text{W}}{\text{VA}_T} = \dfrac{6000}{7200} = 0.83$ or 83%

Power per phase: $P_P = V_P \times I_P \times \text{PF}$
$$= 120\ \text{V} \times 20\ \text{A} \times 0.83$$
$$= 1992\ \text{W}$$

Total power: $P_T = 3 \times V_P \times I_P \times \text{PF}$
$$= 3 \times 120\ \text{V} \times 20\ \text{A} \times 0.83$$
$$= 5976\ \text{W}$$

HIGH-SPEED AND LOW-SPEED GENERATORS

Generators can also be classified as either high-speed or low-speed types (see Figure 4-34). The type of generator used depends on the prime mover or mechanical energy source used to rotate the generator. High-speed generators are usually driven by steam turbines. The generator is smaller in diameter and longer than a low-speed generator. The high-speed generator ordinarily has two stator poles per phase; thus it will rotate at 3600 r/min to produce a 60-Hz frequency.

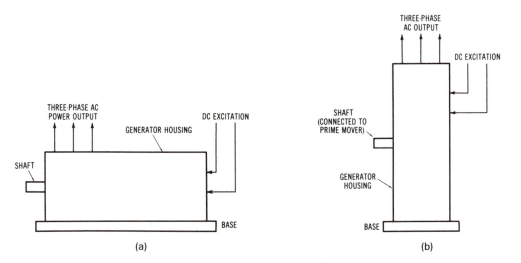

Figure 4-34　Comparison of (a) high-speed and (b) low-speed generator construction.

Low-speed generators are larger in diameter and not as long as high-speed machines. Typical low-speed generators are used at hydroelectric power plants. They have large-diameter revolving fields that use many poles. The number of stator poles used could, for example, be 12 for a 600-r/min machine or eight for a 900-r/min generator. Notice that a much larger number of stator poles is required for low-speed generators.

Generator frequency calculation. The frequency of the sinusoidal wave-forms (sine waves) produced by an ac generator is usually 60 Hz. One cycle of alternating current is generated when a conductor makes one complete revolution past a set of north and south field poles. A speed of 60 r/s (3600 r/min) must be maintained to produce 60 Hz. The frequency of an ac generator (alternator) may be expressed as

$$f = \frac{\text{number of poles per phase} \times \text{speed of rotation (r/min)}}{120} \quad \text{or} \quad \frac{N \times S}{120}$$

where f is the frequency in hertz. Note that if the number of poles is increased, the speed of rotation may be reduced while still maintaining a 60-Hz frequency. The frequency (f) of an ac generator is dependent on the machine's speed of rotation and number of poles per phase.

Assume that a 12-pole three-phase ac generator is rotated at a speed of 1800 r/min. Its frequency is calculated as follows:

$$f = \frac{N \times 5}{120} = \frac{12/3 \times 1800}{120} = \frac{4 \times 1800}{120} = 60 \text{ Hz}$$

The speed of rotation of a generator required for a specific frequency may be found by modifying the frequency formula to solve for speed:

$$\text{speed} = \frac{120 \times \text{frequency}}{\text{poles per phase}}$$

Thus the prime mover speed necessary to generate an ac frequency of 25 Hz for an eight-pole stator is

$$S = \frac{120 \times 25}{8} = 375 \text{ r/min}$$

GENERATOR VOLTAGE REGULATION

As an increased electrical load is added to an alternator, it tends to slow down, due to increased electromagnetic field interaction. The decreased speed causes the generated voltage to decrease since the relative motion between the rotor and the stator is reduced. The amount of voltage change depends on the generator design and the type of load connected to its output terminals. The amount of change in generated voltage from a no-load condition to a rated full-load operating condition is referred to as *voltage regulation*. Voltage regulation may be expressed as

$$\text{VR} = \frac{V_{\text{NL}} - V_{\text{FL}}}{V_{\text{FL}}} \times 100$$

where VR = voltage regulation in percent
       V_{NL} = no-load terminal voltage

V_{FL} = rated full-load terminal voltage

For example, if a generator has a no-load voltage output of 125 V and a rated full-load voltage of 120 V, its voltage regulation is

$$\text{VR} = \frac{V_{\text{NL}} - V_{\text{FL}}}{V_{\text{FL}}} \times 100$$

$$= \frac{125\ \text{V} - 120\ \text{V}}{120\ \text{V}} = \frac{5}{120} \times 100$$

$$= 4.16\%$$

GENERATOR EFFICIENCY

Generator efficiency is the ratio of the power output in watts to the power input in horsepower. The efficiency of a generator may be expressed as

$$\text{efficiency } (\%) = \frac{P_{\text{out}}}{P_{\text{in}}} \times 100$$

where, P_{in} = power input in horsepower
P_{out} = power output in watts

To convert horsepower to watts, remember that 1 hp = 746 W. The efficiency of a generator may range from 70 to 85%.

Assume that a generator has a power output of 5 kW and is driven by an 8-hp prime mover. Its efficiency is

$$\text{efficiency} = \frac{P_{\text{out}}}{P_{\text{in}}} = \frac{5000\ \text{W}}{8 \times 746\ \text{W}} = \frac{5000\ \text{W}}{5968} = 0.837 \quad \text{or} \quad 83.7\%$$

PARALLEL OPERATION OF ALTERNATORS

Alternators for large-scale power generation are operated as several parallel connected units. There are certain advantages in connecting smaller generating units in parallel rather than using one larger unit, even though larger units are more efficient. One advantage of parallel operation is that when one unit is not functional, the system remains in operation at a reduced capacity. Many units may be connected together to supply any conceivable load capacity. Additional units may be added to supply increased load demand. Generating units must be loaded to capacity for maximum efficiency. Thus smaller parallel-connected units may be operated at higher efficiency than a large unit which is underloaded. Maintenance procedures are also simplified by parallel operation since it is feasible to take an alternator "off-line" for repair.

EXAMPLES OF AC GENERATORS

Illustrations of some example types of ac generators or alternators are shown in this section. There are several applications of ac generators other than in power plants for large-scale power generation.

The ac generators shown in Figure 4-35 are called *brushless* generators. These types are typically manufactured in 50- through 1200-kW units. The primary advantage of this type of generator is that it uses an exciter and rotating rectifier unit rather than slip rings and brushes to energize the rotor. Generating units such as this can be used for emergency or standby power generation in buildings or for small-scale power generation for various applications.

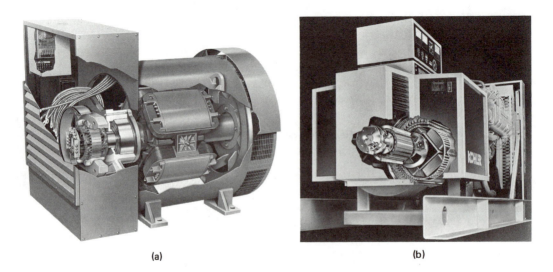

(a)   (b)

Figure 4-35   (a) Brushless ac generator unit; (b) large generator unit with cutaway of rotor and stator assembly. [(a), Courtesy of Marathon Electric; (b), courtesy of Kohler Co.]

Figure 4-36(a) shows a very common type of ac generator. This is a three-phase alternator such as those used in most automobiles today. Automobile alternators generate a three-phase ac voltage which is rectified to produce dc for the charging system. Three-phase alternators are used rather than dc generators since their construction is simpler and maintenance problems are greatly reduced. Figure 4-36(b) shows the stator assembly of an automobile alternator and Figure 4-36(c) shows the rotor assembly. Notice the design features of these assemblies.

In terms of power development, steam turbine-driven ac generators at power plants produce more power than that produced by any other type of generator. The unit shown in Figure 4-37 is a typical steam turbine, three-phase generating unit at a power plant. The unit at the right of the illustration is an exciter unit which contains a rectification system used to develop dc voltage to energize the rotor of the alternator.

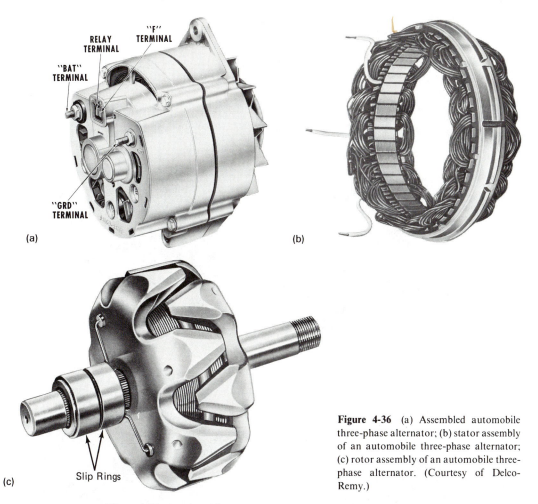

(a)

"BAT"
TERMINAL

RELAY
TERMINAL

"F"
TERMINAL

"GRD"
TERMINAL

(b)

Slip Rings

(c)

Figure 4-36 (a) Assembled automobile three-phase alternator; (b) stator assembly of an automobile three-phase alternator; (c) rotor assembly of an automobile three-phase alternator. (Courtesy of Delco-Remy.)

Figure 4-37 Steam turbine ac generator. (Courtesy of Westinghouse Electric Corp.)

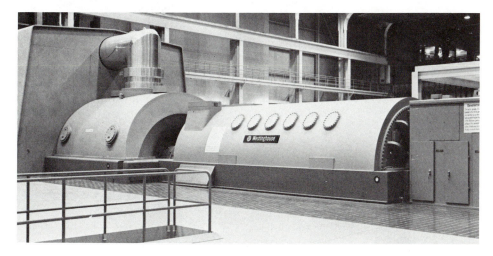

Portable generating units, such as those shown in Figure 4-38, have many applications. They may be used at remote construction sites or for standby power in buildings. Typically, their power rating might be in the range 1200 to 6000 W for lighting, small equipment, and essential appliances or equipment. They may be powered by gasoline or diesel engines or by tractor belt drives.

(a)   (b)

(c)   (d)

Figure 4-38 Portable generator units. [(a)–(c), Courtesy of Ag-tronic, Inc.; (d), courtesy of Kohler Co.]

Diesel electric generator sets, such as the one shown in Figure 4-39, typically come in sizes of 100 to 1000 kW. Units such as this may be used for industries that are remote from utility power lines, for emergency standby power when electricity goes off in a building, for marine applications, or for peak power shaving during daily peak electrical usage. Peak power shaving reduces electrical demand charges for industries and businesses.

Some other portable generator units are shown in Figure 4-40. Portable generator units may use either gasoline, diesel fuel, liquefied petroleum gas, or natural gas as fuel for their prime movers.

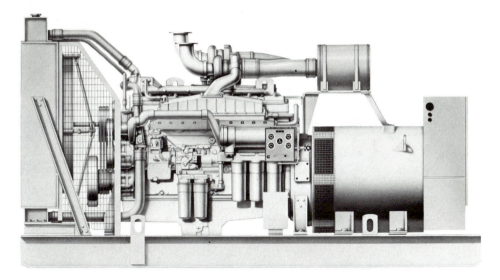

Figure 4-39 Large diesel-engine-driven generator unit. (Courtesy of Cummins Engine Co., Inc.)

(a)　　　　　　　　　　　　　　　　　　　　　　　(b)

Figure 4-40 Generator units. (Courtesy of Kohler Co.)

REVIEW

4.1. What is meant by ac values that are in phase? Out of phase?

4.2. Discuss ac voltage and current characteristics in resistive, inductive, and capacitive circuits.

4.3. Discuss the use of vector or phasor diagrams for ac circuits.

4.4. How is impedance calculated in series ac circuits? In parallel ac circuits?

4.5. Define the following ac power terms: **(a)** apparent power, **(b)** true power, **(c)** reactive power, and **(d)** power factor.

4.6. How does the phase angle of an ac circuit affect the power factor?

4.7. Discuss the basic operational principle of mechanical generators.

4.8. What is the left-hand rule for induced current flow in a generator?

4.9. What three conditions must exist for a mechanical generator to operate?

4.10. Discuss the basic construction features of a mechanical generator.

4.11. Discuss the generation of an ac sine-wave voltage starting with a 0° reference position and progressing through one complete revolution of a conductor.

4.12. How is the instantaneous induced voltage developed in a conductor calculated?

4.13. What are the two types of ac generators?

4.14. Discuss the basic differences between the two types of ac generators.

4.15. What is a general statement of Faraday's law of electromagnetic induction?

4.16. How can the magnitude of instantaneous voltage (V_i) induced into a conductor be calculated?

4.17. What is Lenz's law in relation to ac generators?

4.18. How can average induced voltage (V_{avg}) of an ac generator be calculated?

4.19. Discuss the generation of three-phase ac voltage.

4.20. Discuss the wye and delta methods of connecting the stator windings of a three-phase ac generator.

4.21. What are the advantages of using three-phase ac power?

4.22. Discuss the voltage, current, and power relationships in three-phase wye- and delta-connected ac generators.

4.23. Name some applications of high- and low-speed generators.

4.24. How is the frequency of an ac generator calculated?

4.25. How is the required speed of an ac generator calculated?

4.26. How is voltage regulation of an ac generator calculated?

4.27. How is efficiency of an ac generator calculated?

4.28. What are the advantages of operating alternators in parallel?

4.29. What are some applications of ac generators?

PROBLEMS

4.1. A series ac circuit has 20 V applied, a resistance of 100 Ω, a capacitance of 40 μF, and an inductance of 0.15 H. Calculate:
(a) X_C (b) X_L (c) X_T (d) Z (e) I (f) E_C (g) E_L (h) E_R (i) θ (j) VA (k) W (l) var$_C$ (m) var$_L$ (n) var$_T$ (o) PF

4.2. Draw and label the values of an impedance triangle for Problem 4.1.

4.3. Draw and label the values of a voltage triangle for Problem 4.1.

4.4. Draw and label the values of a power triangle for Problem 4.1.

4.5. Use the same values that were given for R, C, and L in Problem 4.1. Connect them in a parallel circuit with 10 V applied. Calculate:
(a) I_R (b) I_C (c) I_L (d) I_X (e) I_T (f) Z (g) G (h) B_C (i) B_L (j) Y (k) θ (l) VA (m) W (n) var$_C$ (o) var$_L$ (p) PF

4.6. Draw and label the values of a current triangle for Problem 5.5.

4.7. Draw and label the values of an admittance triangle for Problem 5.5.

4.8. Draw and label the values of a power triangle for Problem 5.5.

4.9. An ac circuit converts 12,000 W of power. The applied voltage is 240 V and the current is 72 A. Calculate:
(a) VA **(b)** PF **(c)** θ

4.10. A 240-V single-phase ac generator delivers 20 A of current to a load at a power factor 0.7. Calculate:
(a) Apparent power of the system **(b)** True power **(c)** Reactive power

4.11. The meter connected to a single-phase ac generator indicates the following values:

$$\text{voltage output} = 120 \text{ V}$$
$$\text{current output} = 12 \text{ A}$$
$$\text{power output } = 1 \text{ kW}$$
$$\text{frequency} \quad = 60 \text{ Hz}$$

Calculate:
(a) VA **(b)** PF

4.12. If the peak voltage of a single-phase ac generator is 320 V, what are the instantaneous induced voltages after an armature conductor has rotated the following number of degrees?
(a) 25°   **(b)** 105°
(c) 195°   **(d)** 290°
(e) 325°   **(f)** 350°

4.13. A conductor of an ac generator is 20 in. long and is moved through a uniform magnetic field 42,000 lines/in^2. The distance the conductor travels in 1 s is 400 in. Calculate the instantaneous voltage (V_i) induced into the conductor.

4.14. Calculate the average voltage (V_{avg}) developed in a conductor of an ac generator (one turn) with 8×10^8 magnetic flux lines per pole, rotating at a speed of 3600 r/min ($f = 60$ Hz).

4.15. A three-phase ac wye-connected generator has a phase voltage of 120 V and phase currents of 10 A. Calculate:
(a) Line voltage   **(b)** Line current
(c) Power per phase   **(d)** Total power at 1.0 power factor
(e) Total power at 0.7 power factor

4.16. A three-phase ac delta-connected generator has a line voltage of 480 V, line currents of 7.5 A, and operates at a power factor of 0.85. Calculate:
(a) Phase voltage   **(b)** Phase current
(c) Power per phase   **(d)** Total power

4.17. Calculate the operating frequencies of the following ac generators.
(a) Single-phase, 2-pole, 3600 r/min   **(b)** Single-phase, 4-pole, 1800 r/min
(c) Three-phase, 12-pole, 1800 r/min   **(d)** Three-phase, 6-pole, 3400 r/min

4.18. Calculate the voltage regulation of ac generators with the following no-load and full-load voltage values.
(a) $V_{NL} = 120$ V, $V_{FL} = 115$ V
(b) $V_{NL} = 220$ V, $V_{FL} = 210$ V
(c) $V_{NL} = 480$ V, $V_{FL} = 478$ V

4.19. Calculate the efficiency of the following ac generators.
(a) Power input = 20 hp, power output = 12 kW
(b) Power input = 2.5 hp, power output = 1200 W
(c) Power input = 4.25 hp, power output = 3 kW

FIVE

Direct-Current Generators

Mechanical generators are used in many situations to produce direct-current power. Such generators convert mechanical energy into direct-current (dc) electrical energy. Some examples using dc generators were presented in Chapter 2. These should be reviewed at this time. The parts of a simple dc generator are shown in Figure 5-1.

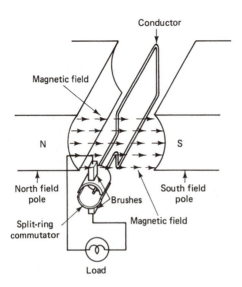

Figure 5-1 Basic parts of a dc generator.

DC GENERATOR OPERATION

The principle of operation of the dc generator is similar to single-phase ac generators discussed in Chapter 4. A rotating armature coil passes through a magnetic field developed between north and south polarities of permanent magnets or electromagnets (see Figure 5-1). As the coil rotates, electromagnetic induction causes a current to be induced into the coil. The current produced is an alternating current; however, it is possible to convert this alternating current into a form of direct current. The conversion of ac to dc is accomplished through the use of a *split-ring commutator*. The split-ring commutator shown in Figure 5-1 has two segments. These segments are insulated from one another and from the shaft of the machine on which the commutator rotates. An end of each armature conductor is connected to each commutator segment. The purpose of the split-ring commutator is to reverse the armature coil connection to the external load circuit at the same time that the current induced in the armature coil reverses. This causes direct current of the correct polarity to be applied to the load circuit at all times.

The voltage developed by the single-coil example used here would appear as illustrated in Figure 5-2. This pulsating direct current is not suitable for most applications. However, by using many turns of wire around the armature, and many split-ring commutator segments, the voltage developed will be a smooth or "pure" direct current such as that produced by a battery. This type of output is shown in Figure 5-3. The voltage developed by a dc generator depends on the strength of the magnetic field, the number of coils in the armature, and the speed of rotation of the armature. By increasing any of these factors, the voltage output of a dc generator can be increased.

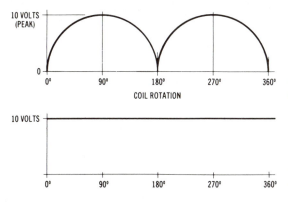

Figure 5-2 Pulsating dc voltage produced by a single-coil dc generator.

Figure 5-3 Pure dc voltage produced by a multicoil dc generator.

DC VOLTAGE GENERATION

The voltage induced into the conductors of a dc generator armature is an ac sine-wave voltage. The sine-wave ac voltage is converted to pulsating dc voltage by the split-ring commutator. One cycle of ac voltage is generated during each 360° rotation of a conductor. The split-ring commutator also serves as a means of connecting the rotating armature conductors to the brushes and external circuit. The brushes of a dc generator are placed directly across from one another. Each brush connects to one side of the commutator to make contact with the armature conductors. The commutator rotates and the brushes are stationary; thus a brush is first in contact with one end of a conductor and then the other (see Figure 5-4). A brush switches from one commutator segment to the other when the conductor reaches a point in its rotation where the induced voltage reverses polarity. This point is called a *neutral plane.* The voltage generated across the brushes changes from minimum to maximum, but always in the same relative direction as shown in Figure 5-4. Figure 5-4 shows five points along the axis of rotation of a conductor within a magnetic field. The position of the brushes in relation to the split-ring commutator is shown in the center of the figure. The lower portion of the figure shows the resulting dc voltage output at each commutator position. Positions 1, 3, and 5 are along the neutral plane (zero induced voltage) of the brush-commutator. Positions 2 and 4 show current flow out of the negative brush, producing a pulsating dc output.

Figure 5-4 Voltage development in a dc generator: (a) rotor conductor placed inside a magnetic field; (b) position of split-ring commutator; (c) resulting dc voltage output at each conductor position.

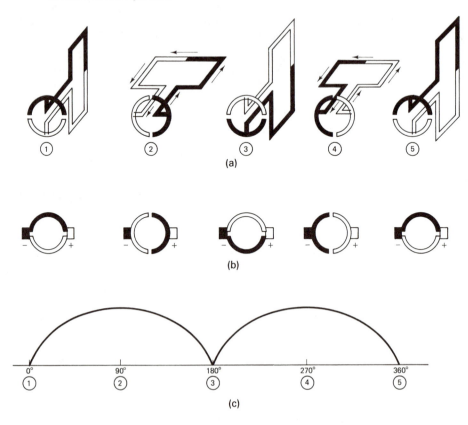

TYPES OF DC GENERATORS

Direct-current generators are classified according to the way in which a magnetic field is developed in the stator of the machine. One method is to use a permanent-magnet field. It is also possible to use electromagnetic coils to develop a magnetic field by applying a separate source of direct current to the electromagnetic coils. However, the most common method of developing a magnetic field is for part of the generator output to be used to supply dc power to the field windings. Thus there are three basic classifications of dc generators: (1) permanent-magnet field, (2) separately excited field, and (3) self-excited field. The self-excited types are further subdivided according to the method used to connect the armature windings to the field circuit. This may be accomplished by the following connection methods: (1) series, (2) parallel (shunt), or (3) compound.

Permanent-magnet dc generator.   A simplified diagram of a permanent-magnet (PM) dc generator is shown in Figure 5-5. The conductors shown in this diagram are connected to the split-ring commutator and brush assembly of the machine. The magnetic field is established by using permanent magnets made of alnico (an alloy of aluminum, nickel, cobalt, and iron, or some other natural magnetic material). It is possible to group several permanent magnets together to create a stronger magnetic field.

The armature of the permanent-magnet dc generator consists of many turns of insulated copper conductors. Therefore, when the armature rotates within the permanent-magnetic field, an induced voltage is developed that can be applied to a

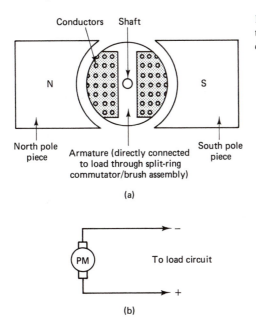

Figure 5-5   Permanent-magnet dc generator: (a) pictorial diagram; (b) schematic diagram.

load circuit. Applications for this type of dc generator are usually confined to those that require low amounts of power. A magneto is an example of a permanent-magnet dc generator.

Separately excited dc generator.   Where large amounts of dc electrical energy are needed, generators with electromagnetic fields are used. Stronger fields can be produced by electromagnets, and it is possible to control the strength of the field by varying the current through the field windings. The output of the generator can thus be controlled with ease.

The direct current used to establish the electromagentic field is referred to as the *exciting current.* When dc exciting current is obtained from a source separate from the generator, it is called a separately excited dc generator. This type of generator is illustrated in Figure 5-6. Storage batteries or rectifiers may be used to supply dc exciting current to this type of generator. The field current is independent of the armature current, and therefore the separately excited generator maintains a very stable output voltage. Changes in load, which affect the armature current, do not vary the strength of the field. The terminal voltage of a separately excited dc generator can be varied by adjusting the current through the field windings. A high-wattage rheostat in series with the field windings will accomplish field control.

Figure 5-6   Separately excited dc generator: (a) pictorial diagram; (b) schematic diagram.

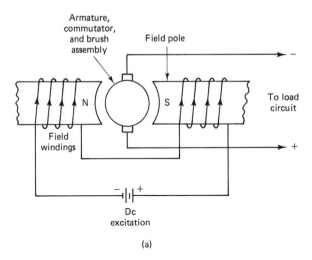

(a)

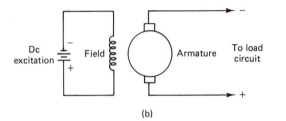

(b)

Separately excited dc generators are used only in certain applications where precise voltage control is essential. Automatic control processes in industry sometimes require such precision. However, the cost of a separately excited dc generator is somewhat prohibitive, and other means of obtaining dc electrical energy are used in certain situations.

Self-excited series-wound dc generator.   Since dc generators produce dc energy, it is possible to extract part of a generator's output to obtain exciting current for the field coils. Generators that use part of their own output to supply exciting current for the electromagnetic field are called *self-excited* dc generators. The method used to connect the armature windings to the field windings determines the characteristics of the generator. It is possible to connect armature and field windings in series, parallel (shunt), or series–parallel (compound).

The self-excited series-wound dc generator has armature windings connected in series with the field windings and the load as shown in Figure 5-7. In this generator, the total current flowing through the load also flows through the field coils. The field coils are therefore wound with low-resistance wire having only a few turns of large-diameter wire. A sufficient electromagnetic field will then be produced by the large current flowing through the low-resistance coils.

When the load is disconnected, no current flows through the generator. However, the field coils retain a small amount of magnetism, known as *residual magnetism.* Due to residual magnetism, a current will begin to flow when a load is

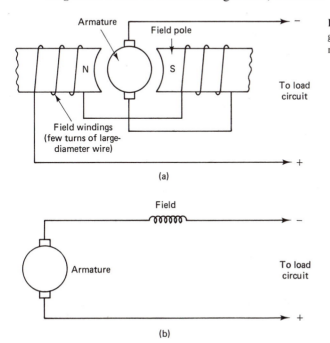

Figure 5-7 Self-excited series-wound dc generator: (a) pictorial diagram; (b) schematic diagram.

connected. The current will then continue to increase, causing a buildup of the magnetic flux of the field. The terminal voltage will rise in proportion to the increases in current. The output curve of a series-wound dc generator is shown in Figure 5-8. When the peak of the curve is reached, magnetic saturation of the field has occurred, prohibiting an increase in terminal voltage. At this point, an increase in load current will cause a rapid decline in terminal voltage due to circuit losses.

The output of a self-excited series-wound dc generator varies appreciably with changes in load current. However, beyond the peak of the output curve, the load current remains fairly stable with large variations in voltage. There are specific applications, such as arc welding, that require stable load current as the voltage changes. However, the self-excited series-wound generator has only a few industrial applications.

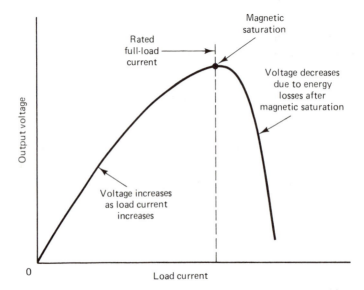

Figure 5-8 Output voltage versus load current curve of a self-excited series-wound dc generator.

Self-excited shunt-wound dc generator. By connecting the field coils, the armature circuit, and a load circuit in parallel, a shunt-wound dc generator configuration is obtained. Figure 5-9 illustrates this type of generator. The output current developed by the generator (I_A) has one path through the (I_L) and another through the field coils (I_F) The generator is usually designed so that the field current is not more than 5% of the total armature current (I_A). The relationship of the currents in a shunt-wound dc generator is $I_A = I_L + I_F$.

In order to establish a strong electromagnetic field and to limit the amount of field current, the field coils are wound with many turns of small-diameter wire. These high-resistance coils develop a strong field due to the large number of turns, and therefore rely less on the amount of field current to develop a magnetic flux.

With no load device connected to the shunt-wound dc generator, a voltage is

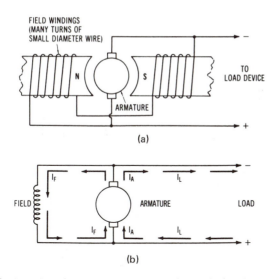

(a)

(b)

Figure 5-9 Self-excited shunt-wound dc generator: (a) pictorial diagram; (b) schematic diagram.

induced as the armature rotates through the electromagnetic field. Again, the presence of residual magnetism in the field coils is critical to the operation of the machine. Current flows in the armature and field circuits as long as there is residual magnetism. As the current increases, the terminal voltage also increases up to a peak level.

When a load device is connected to the generator, the armature current (I_A) increases due to the additional parallel path. This will then increase the $I \times R$ drop in the armature, resulting in a smaller terminal voltage. Further increases in load current cause corresponding decreases in terminal voltage as shown in the output curve of Figure 5-10. With small load currents, the voltage is nearly constant with load current variations. A large load causes the terminal voltage to drop sharply, which is desirable because it provides the generator with a built-in protective feature in case of a short circuit.

The self-excited shunt-wound dc generator is used where a constant output voltage characteristic is needed. It can be used to supply excitation current to a large ac generator, or to charge storage batteries. However, in applications where initial expense is not critical, the compound-wound generator described next may be more desirable.

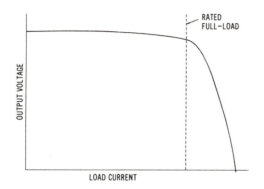

Figure 5-10 Output voltage versus load current curve of a self-excited shunt-wound dc generator.

Self-excited compound-wound dc generator.   The compound-wound dc generator has two sets of field windings. One set is made of low-resistance wire and is connected in series with the armature circuit. The other set is made of high-resistance wire and is connected in parallel with the armature circuit. The compound-wound dc generator is illustrated in Figure 5-11.

As discussed previously, the output voltage of a series-wound dc generator increases with an increase in load current, while the output voltage of a shunt-wound dc generator decreases with an increase in load current. It is possible to produce a dc compound-wound generator, which utilizes both series and shunt windings, that has an almost constant voltage output under changing loads. A constant voltage output may be obtained under varying loads if the series field windings have the proper characteristics to set up a magnetic flux to counterbalance the voltage reduction caused by the $I \times R$ drop in the armature circuit.

A relatively constant output voltage is produced by a *flat-compounded* dc generator. The no-load voltage is equal to the rated full-load voltage in this type of machine, as can be seen in the output curves of Figure 5-12.

Figure 5-11   Self-excited compound-wound dc generator: (a) pictorial diagram; (b) schematic diagrams.

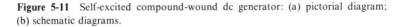

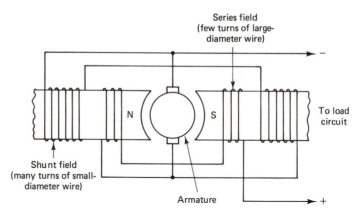

(a)

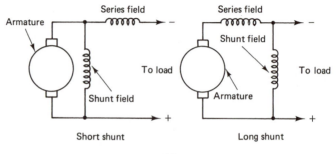

(b)

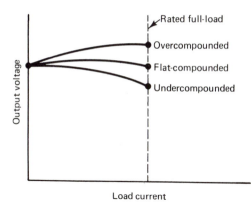

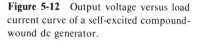

Figure 5-12 Output voltage versus load current curve of a self-excited compound-wound dc generator.

If the series field has more electromagnetic field strength, the generator will possess a series characteristic. The voltage output will increase with an increase in load current. A compound-wound generator whose full-load voltage is greater than its no-load voltage is called an *overcompounded* dc generator. If the shunt field has more electromagnetic field strength, the output will be more characteristic of a shunt generator. Such a generator, whose full-load voltage is less than its no-load voltage, is called an *undercompounded* dc generator.

Dc compound-wound generators can be constructed so that the series and shunt fields either aid or oppose one another. If the magnetic polarities of adjacent fields are the same, the magnetic fluxes aid each other and are said to be *cumulatively* wound. Opposing polarities of adjacent coils produce a *differentially* wound machine. For almost all applications of compound-wound machines, the cumulative method is used. In this way, a generator wound maintains a fairly constant voltage output with variations in load current. The compound-wound dc generator is used more extensively than other dc generators due to its constant voltage output and its design flexibility to obtain various output characteristics.

VOLTAGE BUILDUP IN SELF-EXCITED DC GENERATORS

Self-excited dc generators develop their own dc excitation voltage for the field windings. Initial voltage buildup is dependent primarily on residual magnetism in the field windings. When a self-excited dc generator starts from 0 r/min, the generated voltage (V_g) is zero. As prime mover speed increases, the generated voltage due to the presence of residual magnetism and relative speed increases also. The current flow due to increased generated voltage increases, causing an increase in magnetomotive force (MMF) around the field coils. The MMF aids the residual magnetism and causes V_g to increase in direct proportion. This effect is cumulative and results in rated voltage output at rated prime mover speed.

If voltage buildup does not occur, there are several conditions that should be checked. The first reason for lack of voltage buildup is low residual magnetism in the field windings. This condition may be remedied by "flashing the field" (applying dc voltage to the field circuit and then removing it). This procedure should restore residual magnetism to the field coils. A second condition that may occur is the reverse of field circuit connections with respect to armature connections. To remedy this situation, either the field circuit *or* armature circuit connections may be reversed or the prime mover direction of rotation can be reversed to accomplish the same purpose. Another factor that can cause failure of a dc self-excited generator to build up to rated voltage output is excessively high resistance in either the field or armature circuit. Either situation causes reduced current flow which prohibits voltage development.

CURRENT AND VOLTAGE RELATIONSHIPS IN DC GENERATORS

Series-wound dc generators have current and voltage characteristics similar to those of basic series dc circuits. The current flow through a series-wound dc generator is the same through each part of the circuit. Thus $I_A = I_F = I_L$, where I_A is the induced armature current, I_F the field current, and I_L the current flow through the load circuit. The voltage distribution of a series-wound dc generator is as follows: $V_A = V_L + V_F$, where V_A is the voltage across the armature, V_L the voltage across the load circuit, and V_F the voltage across the field windings. Also, it should be noted that $V_A = V_g - I_A R_A$, where V_g is the voltage generated in the armature and $I_A R_A$ is the voltage drop across the armature due to the armature current (I_A) through a fixed resistance (R_A). In addition, the voltage across the field windings (V_F) is equal to $I_F R_F$, where I_F is the field current and R_F is the field resistance.

  Shunt-wound dc generators are similar to basic parallel dc circuits in characteristics. The current distribution is $I_A = I_F + I_L$, where I_F is field current and equals V_F / R_F and I_L is load current and equals V_L / R_L. The voltage is the same across each path of a parallel circuit. Thus $V_A = V_F = V_L$, where $V_A = V_g - I_A R_A$.

  Compound-wound dc generators have two methods of winding connection: the long-shunt and short-shunt, as shown in Figure 5-11(b). The current distribution in the *long-shunt* connection is $I_A = I_F + I_L = I_S$, where I_S is the current flow through the series field. The voltages of a long-shunt generator are distributed as $V_A = V_S + V_F$, where V_S is the voltage across the series field and $V_F = V_L$. The *short-shunt* connection has current flows distributed as $I_A = I_F + I_S$ and $I_S = I_L$. The voltage relationship of the short-shunt dc generator connection is $V_A = V_F = V_S + V_L$.

COMMON DC GENERATOR CHARACTERISTICS

Dc generators are used primarily for operation with mobile equipment. In industrial plants, they are used for standby power, battery charging, and for specialized dc operations such as electroplating. In many situations, rectification systems such as the one shown in Figure 5-13, which convert alternating current to direct current, have replaced dc generators. For most applications, they are cheaper to operate and maintain.

Figure 5-13 Industrial direct-current rectification system. (Courtesy of Kinetics Industries.)

Dc generators supply energy by converting the mechanical energy of some prime mover, such as a gasoline or diesel engine, into electrical energy. The prime mover must rotate at a definite speed.

Armature reaction. When dc generators operate, a characteristic known as *armature reaction* takes place. The current through the armature windings produces a magnetix flux, which reacts with the main field flux as shown in Figure 5-14. The result

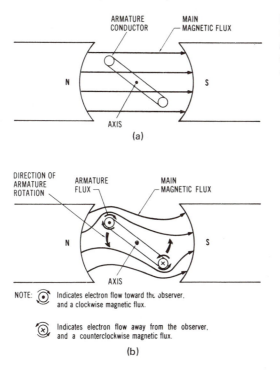

Figure 5-14 Armature reaction: (a) main magnetic flux with no current in the armature windings; (b) distortion of main magnetic flux with current in the armature windings.

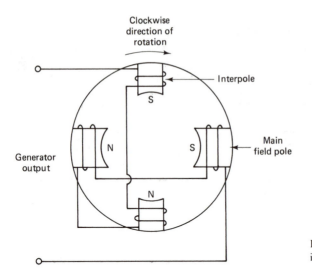

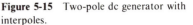

Figure 5-15   Two-pole dc generator with interpoles.

is that a force is created which tends to rotate the armature in the direction indicated. As the load current increases, the increase in armature current causes a greater amount of armature reaction to take place. This condition can cause a considerable amount of sparking between the brushes and the commutator. However, armature reaction can be reduced by placing windings called *interpoles* between the main field windings of the stator. These windings are connected in series with the armature windings. Thus an increase in armature current creates a stronger magnetic flux around the interpole, which counteracts the main field distortion created by the armature conductors. A two-pole generator with interpoles is illustrated in Figure 5-15.

Compensating windings.   Another method of reducing the effect of armature reaction is to design a dc generator with *compensating windings*. These are small windings placed inside the main field poles and connected in series with the armature windings. The current flow and resulting magnetic field around the compensating windings is opposite that of adjacent armature coils, as shown in Figure

Figure 5-16   Compensating windings to reduce armature reaction.

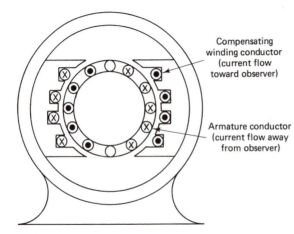

Compensating winding conductor (current flow toward observer)

Armature conductor (current flow away from observer)

5-16. The magnetic fields developed around the compensating windings tend to counteract the fields developed around the main field windings and reduce armature reaction.

Dc generator ratings.   Dc generator output is usually rated in kilowatts, which is the rated electrical power capacity of a machine. Other ratings, which are specified by the manufacturer on the nameplate of the machine, are current capacity, terminal voltage, speed, and temperature. Dc generators are made in a wide range of physical sizes, with various electrical characteristics.

ROTARY CONVERTERS

A type of rotating machine used to convert alternating current to direct current is called a rotary converter. Rotary ac-to-dc converters are used for limited applications. A motor-driven generator unit is illustrated in Figure 5-17. Motor-generator units may also be used to convert direct current to alternating current. This function is referred to as an *inverter.* When operated as a converter to produce dc, the machine is run off an ac line. This process is shown in Figure 5-17. The ac is transferred to the machine windings through slip rings, and converted to dc by a split-ring commutator located on the same shaft. The amount of dc voltage output is determined by the magnitude of the ac voltage applied to the machine and the winding ratio of the rotor. Converters may

Figure 5-17   Rotary ac-to-dc converter: (a) block diagram; (b) schematic diagram.

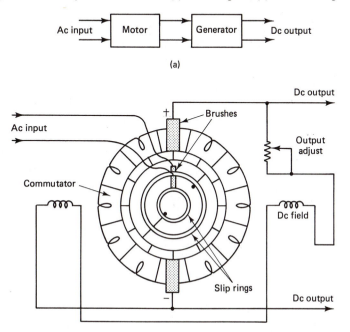

(a)

(b)

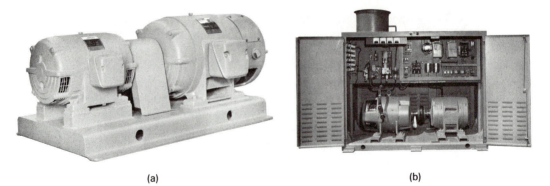

(a)   (b)

Figure 5-18 Typical motor-generator assemblies. [(a), Courtesy of Litton Industrial Products, Inc., Louis-Allis Division; (b), courtesy of Kinetics Industries.]

be designed as two units, with motor and generator shafts coupled together or as one unit housing both the motor and generator. Two typical motor-generator assemblies are shown in Figure 5-18.

REVIEW

5.1. What is the purpose of the split-ring commutator of a dc generator?

5.2. Discuss the production of a pulsating dc voltage in a conductor rotated inside a magnetic field.

5.3. How can the voltage output of a dc generator be increased?

5.4. What are the three basic classifications of dc generators?

5.5. Discuss the operation of the following types of dc generators: **(a)** permanent magnet, **(b)** separately excited, **(c)** series-wound, **(d)** shunt-wound, and **(e)** compound-wound.

5.6. What is meant by the following terms relating to dc compound-wound generators: **(a)** flat-compounded, **(b)** overcompounded, **(c)** undercompounded, **(d)** cumulatively wound, **(e)** differentially wound, **(f)** short-shunt connection, and **(g)** long-shunt connection?

5.7. Discuss voltage buildup in dc generators.

5.8. What is meant by "flashing the field" of a dc generator?

5.9. Discuss current and voltage relationships in the following dc generators: **(a)** series-wound, **(b)** shunt-wound, **(c)** compound-wound long-shunt, and **(d)** compound-wound short-shunt.

5.10. What are some applications of dc generators.

5.11. Discuss armature reaction in dc generators.

5.12. What are interpoles? What are compensating windings?

5.13. How are dc generators rated?

5.14. What is a rotary ac-to-dc converter?

5.15. What is an inverter?

PROBLEMS

5.1. A series-wound dc generator has an armature resistance of 1.5 Ω and a field resistance of 4.5 Ω. The voltage across a 12-Ω load is 100 V. What are the values of:
(a) Armature current? (b) Power output?

5.2. A shunt-wound dc generator has an armature resistance of 2.5 Ω and a field resistance of 40 Ω. Its output voltage is 120 V. Calculate:
(a) Armature current   (b) Field current
(c) Load current   (d) Power output

5.3. A 120-V dc generator is rated at 20 kW. What is its maximum load current?

5.4. A 100-kW, 240-V shunt-wound dc generator has a field resistance of 60 Ω and an armature resistance of 0.1 Ω. Calculate:
(a) Rated load current   (b) Field current
(c) Armature current   (d) Generated voltage (V_g) = output voltage: $I_A \times R_A$

5.5. A long-shunt compound-wound dc generator is rated at 150 kW and 480 V. Its armature resistance is 0.05 Ω, shunt-field resistance 100 Ω, and series field resistance 0.02 Ω. Calculate:
(a) Rated load current (b) Field current (c) Armature current

5.6. The no-load voltage of a dc generator is 130 V and the full-load voltage is 120 V. Calculate the voltage regulation (%) of the generator.

5.7. A 50-kW 240-V shunt-wound dc generator produced a generator voltage (V_g) of 255 V. Its field current is 2 A. Calculate:
(a) Field resistance (b) Armature resistance

5.8. A 2-kW 100-V series-wound dc generator has an armature resistance of 2 Ω and a field resistance of 5 Ω. Calculate:
(a) Field current (b) Armature current (c) Generated voltage

SIX

Transformers

The heart of power distribution systems is a device known as a *transformer*. This device is capable of controlling massive amounts of power for efficient distribution. Transformers are used for many applications, as shown in Figure 6-1.

Figure 6-1 (a) Small pulse transformers; (b) small transformers for printed circuit boards; (c) current transformer. [(a), (b), Courtesy of TRW Inductive Products Division; (c), courtesy of Basler Electric Company.]

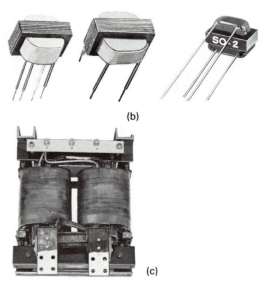

(a) (b) (c)

TRANSFORMER OPERATION

Transformers provide a means of converting an ac voltage from one value to another. The basic construction of a transformer is illustrated in Figure 6-2. Notice that the transformer consists of two sets of windings that are not physically connected. The only connection between the primary and secondary windings is the magnetic coupling effect, known as *mutual induction,* which takes place when the circuit is energized by an ac source. The metallic core plays an important role in transferring magnetic flux from the primary winding to the secondary winding.

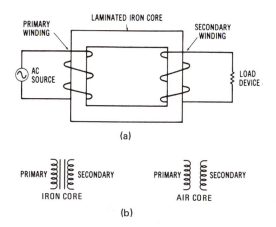

(a)

(b)

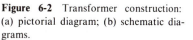

Figure 6-2 Transformer construction: (a) pictorial diagram; (b) schematic diagrams.

If an alternating current, which is constantly changing in value, flows in the primary winding circuit, the magnetic flux produced around the primary will be transferred to the secondary winding. Thus an induced voltage is developed across the secondary winding. In this way, electrical energy is transferred from the source to the load circuit.

The efficient transfer of energy from primary windings to secondary windings depends on the flux linkage of the magnetic field between these sets of windings. Ideally, all magnetic flux lines developed around the primary would be transferred by flux linkage to the secondary. However, a certain amount of flux leakage takes place as some lines of force escape to the surrounding air.

To decrease the amount of flux leakage, transformer windings are wound around laminated iron cores. Iron cores concentrate the magnetic flux so that better flux linkage between primary and secondary is accomplished. Two types of cores are illustrated in Figure 6-3. These cores are made of laminated iron to reduce undesirable eddy currents, which are induced into the core material. Thus the laminations reduce heat loss. The diagram of Figure 6-3(a) shows closed-core transformer construction. The windings are placed along the outside of the metal core. Figure 6-3(b) shows the shell-core construction. This type produces better magnetic coupling since the windings are surrounded by metal on both sides. Note that the primary and secondary windings are placed adjacent to one another for better coupling.

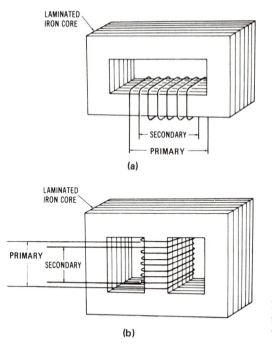

(a)

(b)

Figure 6-3 Types of iron core transformers: (a) closed-core type; (b) shell-core type.

Transformer efficiency.   Transformers are very efficient electrical devices. A typical efficiency rating is around 98%. Transformer efficiency is expressed as

$$\text{efficiency (\%)} = \frac{P_{\text{out}}}{P_{\text{in}}} \times 100$$

where P_{out} = power output in watts
      P_{in} = power input in watts

For example, assume that a transformer has a primary voltage of 120 V, a primary current of 2 A, a secondary voltage of 240 V, and a secondary current of 0.96 A. The efficiency of the transformer is found as follows:

primary power: $P_{\text{pri}} = V_{\text{pri}} \times I_{\text{pri}}$
                    $= 120\text{V} \times 2\text{A}$
                    $= 240 \text{ W}$

secondary power: $P_{\text{sec}} = V_{\text{sec}} \times I_{\text{sec}}$
                    $= 240\text{V} \times 0.96\text{A}$
                    $= 230.4 \text{ W}$

efficiency: $\dfrac{P_{\text{out}}}{P_{\text{in}}} = \dfrac{P_{\text{sec}}}{P_{\text{pri}}} = \dfrac{230.4}{240 \text{ W}} = 0.96 \quad \text{or} \quad 96\%$

The losses that reduce efficiency, in addition to flux leakage, are copper and iron losses. Copper loss is the heat (I^2R) loss of the windings, while iron losses are those caused by the metallic core material. The insulated laminations of the iron core help to reduce iron losses.

Step-up and step-down transformers. Transformers are functionally classified as step-up or step-down types. These types are illustrated in Figure 6-4.

The step-up transformer of Figure 6-4(a) has fewer turns of wire on the primary than on the secondary. If the primary winding has 50 turns of wire and the secondary has 500 turns, a turns ratio of 1:10 is developed. Therefore, if 12 V ac is applied to the primary from the source, 10 times that voltage, or 120 V ac, will be transferred to the secondary load circuit (assuming no losses).

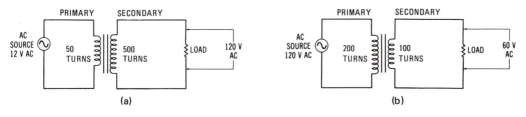

Figure 6-4 Transformers: (a) step-up type; (b) step-down type.

The example of Figure 6-4(b) is a step-down transformer. The step-down transformer has more turns of wire on the primary winding than on the secondary winding. The primary winding of the example has 200 turns while the secondary winding has 100 turns, or a 2:1 ratio. If 120 V ac is applied to the primary from the source, then one-half that amount, or 60 V ac, will be transferred to the secondary load.

Transformer voltage and current ratios. The preceding examples show that there is a direct relationship between the primary and secondary turns of wire and the voltages across each winding. The voltage ratio can be expressed as

$$\frac{V_P}{V_S} = \frac{N_P}{N_S}$$

where V_P = voltage across the primary winding
V_S = voltage across the secondary winding
N_P = number of turns in the primary winding
N_S = number of turns in the secondary winding

The transformer is a power-control device; therefore, the following relationship can be expressed:

$$P_P = P_S + \text{losses}$$

where P_P = primary power
P_S = secondary power

and the losses are those that ordinarily occur in a transformer. In transformer theory, an ideal device is usually assumed, and losses are not considered. Thus, since $P_P = P_S$ and $P = V \times I$, then

$$V_P \times I_P = V_S \times I_S$$

where I_P = primary current
I_S = secondary current

Therefore, if the voltage across the secondary is stepped up to twice the voltage across the primary, the secondary current will be stepped down to one-half the primary current. The current ratio of a transformer is thus expressed as

$$\frac{I_P}{I_S} = \frac{N_S}{N_P}$$

Note that whereas the voltage ratio is a *direct* relationship, the current ratio is an *inverse* relationship.

Multiple-secondary transformer. It is also possible to construct a transformer that has multiple secondary windings, as shown in Figure 6-5. The transformer in this example is connected to a source of 120 V ac, which produces the primary magnetic flux. The secondary has two step-down windings and one step-up winding. Between points 1 and 2, a voltage of 6.3 V ac is developed. Between points 5 and 6, a voltage of 30 V ac is obtained and between points 3 and 4, a voltage of 360 V ac is developed. This type of transformer is often used for the power supply circuit of various types of electronic equipment and instruments.

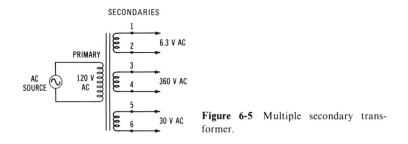

Figure 6-5 Multiple secondary transformer.

Autotransformers. Another specialized type of transformer is the autotransformer, illustrated in Figure 6-6. The autotransformer has only one winding, with a common connection between the primary and secondary windings. The principle of operation of the autotransformer is similar to that of other transformers. Both the step-up and step-down types are shown in Figure 6-6(a). Figure 6-6(b) shows a variable autotransformer in which the winding tap can be adjusted along the entire length of the winding to provide a variable ac voltage to a load.

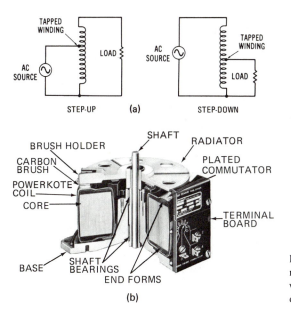

Figure 6-6 Autotransformers: (a) schematic diagrams; (b) cutaway view of a variable autotransformer. [(b), Courtesy of Superior Electric Co.]

POWER DISTRIBUTION TRANSFORMERS

The distribution of electrical power in the United States is normally three-phase 60-Hz alternating current. This power, of course, can be manipulated or changed in many ways by the use of electrical circuitry. For instance, a rectification system is capable of converting three-phase 60-Hz alternating current to a form of direct current, as shown in Figure 6-7. Single-phase power is generally suitable for lighting and small appliances such as those used in the home. However, three-phase power is more economical where a large amount of electrical energy is required. Since industries use over 40% of all electrical power produced, three-phase power is distributed directly to most industries. Electrical substations located near industries employ massive power distribution transformers and associated equipment such as oil-filled circuit breakers, high-voltage conductors, and huge strings of insulators to distribute power to industry. From these substations, power is brought into an industrial site so that the many kinds of equipment used for mass production and specialized industrial processes can be energized. Transformers play a critical role in ac power distribution.

The distribution of ac power is dependent on the use of transformers at many points along the power distribution system. To transmit electrical power over long distances, less current is required at high voltages, since *line loss* (I^2R) is reduced significantly. A typical high-voltage transmission line may extend a distance of 50 to 100 miles from the generating station to the first substation. These high-voltage power transmission lines may operate at 300,000 to 500,000 V by using step-up transformers

Figure 6-7 Three-phase rectification system—converts three-phase ac to dc. (Courtesy of Kinetics Industries.)

to increase the voltage produced by the ac generators at the power station. Various substations are encountered along the power distribution system, where transformers are used to reduce the high transmission voltages to a voltage level such as 480 V, which is suitable for industrial motor loads, or 120/240 V for residential use. Power distribution systems are discussed in detail in Chapter 7.

Transformer polarity.   Power distribution transformers usually have polarity markings so that their windings can be connected in parallel to increase their current capacity. The standard markings are H_1, H_2, H_3, and so on, for the primary windings, and X_1, X_2, X_3, and so on, for the secondary windings. Many power transformers have two similar primary windings and two similar secondary windings to make them adaptable to different voltage requirements simply by changing from a series to a parallel connection. The voltage combinations available from this type of transformer are shown in Figure 6-8.

Transformer ratings.   The ratings of power transformers are very critical. Usually, transformers are rated in kilovoltamperes (kVA). A kilowatt rating is not used since it would be misleading due to the various power-factor ratings of industrial loads. Other power transformer ratings usually include frequency, rated voltage of each winding, and an impedance rating.

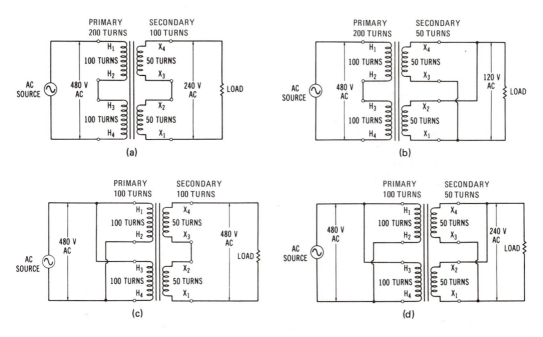

Figure 6-8 Transformer connection methods for various voltage combinations using a distribution transformer with a 2:1 turns ratio: (a) series-connected primary and secondary; (b) series-connected primary, parallel-connected secondary; (c) parallel-connected primary, series-connected secondary; (d) parallel-connected primary and secondary.

Transformer cooling. Power transformers located along the power distribution system operate at very high temperatures. The purpose of the cooling equipment is to conduct heat away from the transformer windings. Many power transformers are of the liquid-immersed type. The windings and core of the transformer are immersed in an insulating liquid contained in the transformer enclosure. The liquid insulates the windings as well as conducts heat away from them. Askarel is an insulating liquid that is used extensively. Some transformers, called dry types, use forced air or inert gas as coolants. Some locations, particularly indoors, are considered hazardous for the use of liquid immersed transformers. However, most transformers rated at over 500 kVA are liquid filled.

THREE-PHASE POWER TRANSFORMERS

Since industries predominately use three-phase power, they rely on three-phase distribution transformers, such as the one shown in Figure 6-9, to supply this power. Usually, when single-phase voltage is required, it is taken from three-phase power lines. Large three-phase distribution transformers are usually located at substations adjoining industrial plants. Their purpose is to supply the proper ac voltages to meet industrial load requirements. The ac voltages transmitted to industrial substations are high voltages that must be stepped down by three-phase distribution transformers.

Figure 6-9 Three-phase transformer.
(Courtesy of Basler Electric Company.)

Three-phase transformer connections.   There are five common methods that can be used to connect the primary and secondary windings of three-phase transformers. These methods are the (1) delta-delta, (2) delta-wye, (3) wye-wye, (4) wye-delta, and (5) open delta. These basic methods are illustrated in Figure 6-10. The delta-delta connection is used for some lower-voltage applications. The delta-wye method is commonly used for stepping voltages up, since an inherent step-up factor of 1.73 times is obtained due to the voltage characteristic of the wye-connected secondary. The wye-wye connection is ordinarily not used, while the wye-delta method may be advantageously used to step voltages down. The open-delta connection is used if one transformer winding becomes damaged or is taken out of service. The transformer will still deliver three-phase power, but at a lower current and power capacity. This connection may also be desirable where the full capacity of three transformers is not needed until a later time. Two identical single-phase transformers could be used to supply power to the load until, at a later time, the third transformer is needed to meet increased load requirements.

Three-phase distribution systems.   Three-phase power distribution systems that supply industries may be classified according to the number of phases and number of wires required. These systems, shown in Figure 6-11, are the (a) three-phase three-wire system; (b) three-phase three-wire system with neutral; and (c) three-phase four-wire system. The primary winding connection method is not considered here. The three-phase three-wire system can be used to supply motor loads of 240 or 480 V, for instance. Its major disadvantage is that it only supplies one voltage. Only three hot lines are supplied to the load. The usual insulation color code on these hot lines is black, red, or blue, as specified in the *National Electrical Code®*.

The disadvantage of the three-phase three-wire system can be partially overcome by adding one center-tapped phase winding as shown in the three-phase, three-wire with neutral system in Figure 6-11(b). This system could be used as a supply for

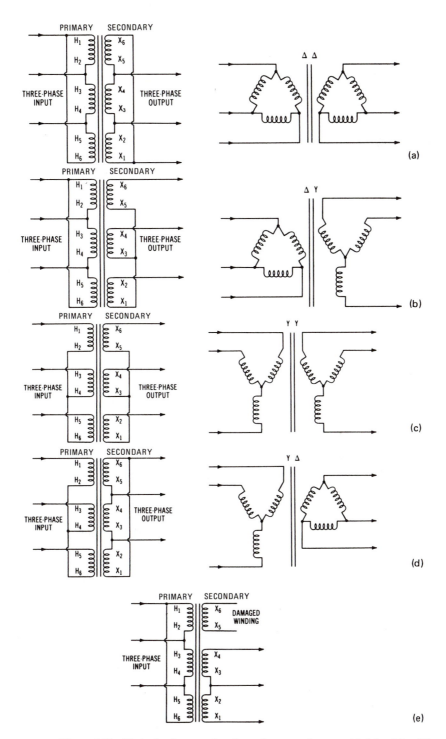

Figure 6-10 Methods of connecting three-phase transformers: (a) delta-delta; (b) delta-wye; (c) wye-wye; (d) wye-delta; (e) open delta.

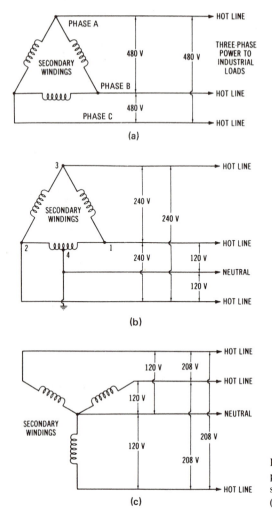

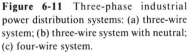

Figure 6-11 Three-phase industrial power distribution systems: (a) three-wire system; (b) three-wire system with neutral; (c) four-wire system.

120/240 or 240/480 V. Assume that it is used to supply 120/240 V. The voltage from the hot line at point 1 and the hot line at point 2 to neutral would be 120 V, due to the center-tapped winding. Across any two hot lines, 240 V would still be available. The neutral wire is color coded with white or gray insulation. The disadvantage of this system is that when making wiring changes, it would be possible to connect a 120-V load between the neutral and point 3 (sometimes called the "wild" phase). The voltage present here would be the combination of three-phase voltages between points 1 and 4 and points 1 and 3. This voltage would be in excess of 200 V! Although the wild-phase situation exists, this system is capable of supplying both high-power loads and low-voltage loads such as lighting and small equipment.

The most widely used industrial distribution system is the three-phase four-wire system. This system, shown in Figure 6-11(c), commonly supplies 120/208 and 277/480 V for industrial load requirements. The 120/208-V system is illustrated here. From neutral to any hot line, 120 V for lighting and low-power loads may be obtained. Across any two hot lines, 208 V is present for supplying motors or other high-power loads. The most popular system for industrial and commercial power distribution is the 277/408-V system, which is capable of supplying both three-phase and single-phase loads. A 240/416-V system is also used for industrial loads, while the 120/208-V system is often used for underground distribution in urban areas. Note that this system is based on the voltage characteristics of the three-phase wye connection and that the relationship $V_L = V_P \times 1.73$ exists for each application of this system.

SINGLE-PHASE POWER TRANSFORMERS

Single-phase distribution systems can be considered in the same way as three-phase systems. Although single-phase systems are used mainly for residential power distribution systems, there are some industrial applications of single-phase systems. Single-phase power distribution is usually originated from three-phase power lines. It is possible to obtain three separate single-phase voltages from a three-phase line. Thus our power systems are capable of supplying both three-phase and single-phase loads from the same power lines. Figure 6-12 shows a typical power distribution system from the power station (source) to the various loads that are connected to the transformers.

Single-phase systems could be of two major types: (1) single-phase two-wire systems and (2) single-phase three-wire systems. A single-phase two-wire system is shown in Figure 6-13(a). This system uses a transformer whose secondary produces one single-phase voltage, such as 120 or 240 V. In early residential distribution systems, this type was the most used to provide 120-V service. However, as appliance power requirements increased, the need for a dual-voltage system was evident.

To meet the demand for more residential power, the single-phase three-wire system is now used. A home service entrance could be supplied with 120/240-V electrical energy by the system of Figure 6-13(b) or (c). Each of these systems is derived from a three-phase power line. The single-phase three-wire system has two hot lines and a neutral. The hot lines (black and red insulated wire) are connected to the ends of the transformer secondary windings. The neutral (white insulated wire) is connected to the center tap of the distribution transformer. Thus, from neutral to either hot line, 120 V for lighting and low-power requirements is available. Across the hot lines, 240 V is supplied for high-power requirements. Therefore, the current requirement for large-power-consuming equipment is cut in half, since 240 V rather than 120 V is used. Either the single-phase two-wire system or the single-phase three-wire system could be used to supply single-phase power to industry. However, these single-phase systems are used mainly for residential power requirements.

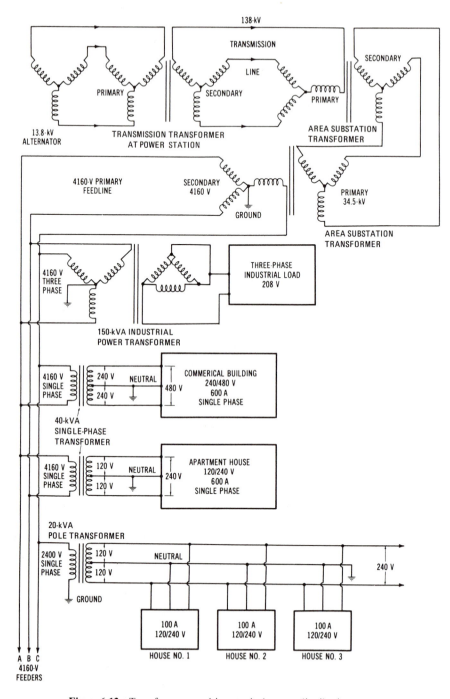

Figure 6-12 Transformers used in a typical power distribution system.

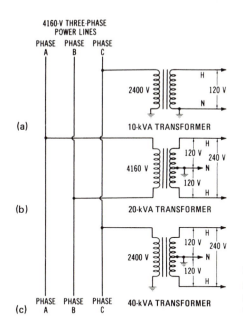

Figure 6-13 Transformers used for single-phase power distribution: (a) two-wire system; (b) three-wire system; (c) three-wire system.

TRANSFORMER MALFUNCTIONS

Transformer malfunctions result when a circuit problem causes the insulation to break down. Insulation breakdown permits electrical arcs to flow from one winding to an adjacent winding. These arcs, which may be developed throughout the transformer, cause a decomposition of the winding insulation. This can be a particularly hazardous problem for larger power transformers, since a gas may be produced due to the reaction of the electric arc and the insulation. For this reason, it is very important for circuit protection to be provided for transformers. They should have power removed promptly whenever some type of fault develops. Current-limiting fuses may also be used to respond rapidly to a circuit malfunction.

REVIEW

6.1. What are the basic parts of a transformer?

6.2. Discuss the operation of a transformer.

6.3. What is the purpose of using a laminated metal transformer core?

6.4. Discuss the two types of core construction used for transformers.

6.5. How is transformer efficiency calculated?

6.6. What is a step-up transformer? A step-down transformer?

6.7. What is the voltage ratio of a transformer?

6.8. What is the current ratio of a transformer?

6.9. What is the relationship of primary and secondary power in an "ideal" transformer?

6.10. What are the following types of transformers: **(a)** multiple secondary, **(b)** autotransformer, and **(c)** variable autotransformer?

6.11. Why is ac power distributed over long distances rather than dc?

6.12. Why is transformer polarity important?

6.13. What type of power rating is used for transformers?

6.14. What types of cooling methods are used for power transformers?

6.15. What are the five methods used to connect the primary and secondary windings of three-phase transformers?

6.16. What are the three classifications of three-phase power distribution systems?

6.17. Discuss the three-phase four-wire system used for industrial power distribution.

6.18. What are the two types of single-phase distribution systems?

6.19. Discuss the single-phase three-wire system used for residential power distribution.

6.20. What ordinarily causes a transformer malfunction?

PROBLEMS

6.1. A 1500-turn primary winding has 120 V applied. The secondary voltage is 20 V. How many turns of wire does the secondary winding have?

6.2. When a 200-Ω resistance is placed across a 240-V secondary winding the primary current is 30 A. What is the value of the primary voltage of the transformer?

6.3. A transformer has a 20:1 turns ratio. The primary voltage is 4800 V. What is the secondary voltage?

6.4. A transformer has a 2400-V primary winding and a 120-V secondary. The primary winding has 1000 turns. How many turns of wire does the secondary winding have?

6.5. A 240- to 4800-V transformer has a primary current of 95 A and a secondary current of 4 A. What is its efficiency?

6.6. A transformer has a 20-kVA power rating. Its primary voltage is 120 V and its secondary voltage is 480 V. What are the values of its maximum:
(a) Primary current?   **(b)** Secondary current?

6.7. A single-phase transformer with two primary and two secondary windings with an equal number of turns is connected as follows to a 480-V source. Primary windings are in series and secondary windings are in parallel. What is the secondary voltage?

6.8. A three-phase transformer is connected in a wye-delta configuration. The primary voltage is 4800 V, the primary current is 100 A, and the turns ratio is 10:1. Calculate:
(a) Secondary voltage   **(b)** Secondary current
(c) Primary power   **(d)** Secondary power

6.9. A three-phase transformer is connected in a delta-wye configuration. The primary voltage is 4160 V, the primary current is 40 A, and the turns ratio is 10:1. Calculate:
(a) V_S **(b)** I_S
(c) P_P **(d)** P_S

SEVEN

Power Distribution Systems

Power distribution systems are used to transfer electrical power from an alternating-current or a direct-current source. Distribution networks to distribute power from where it is produced to where it is used can be quite simple. For example, a battery could be connected directly to a small motor, with only a set of wires to connect them together.

More complex power distribution systems are used, however, to transfer electrical power from a power plant to industries, homes, and commercial buildings. Distribution systems usually employ transformers, conductors, and protective devices as a minimum.

POWER TRANSMISSION AND DISTRIBUTION

The distribution of electrical power in the United States is normally in the form of three-phase 60-Hz alternating current. This power, of course, can be manipulated or changed in many ways by the use of electrical circuitry. Single-phase power is generally used for lighting and small appliances in a residential environment. However, where a large amount of electrical power is required, such as for industrial and commercial buildings, three-phase power is more economical.

The distribution of electrical power involves a very complex system of interconnected power transmission lines. These transmission lines originate at the electrical power-generating stations located throughout the United States. The ultimate purpose of these power transmission and distribution systems is to supply the electrical power necessary for industrial, residential, and commercial use.

Figure 7-1 Electrical power distribution substation. (Courtesy of Kuhlman Corp.)

Industries use almost 50% of all the electrical power produced, so three-phase power is distributed directly to most industries. Substations such as the one shown in Figure 7-1 use massive transformers and associated equipment, such as oil-filled circuit breakers, high-voltage conductors, and huge strings of insulators for distributing electrical power. From these substations, power is distributed to energize industrial machinery, homes, and commercial buildings.

Power transmission and distribution systems are used to interconnect electrical power production systems and to provide a means of delivering electrical power from the generating station to its point of utilization. Most electrical power systems east of the Rocky Mountains are interconnected with one another in a parallel circuit arrangement. These interconnections of power production systems are monitored and controlled, in most cases, by a computerized control center. Such control centers provide a means of data collection and recording, system monitoring, frequency control, and signaling. Computers have become an important means of assuring the efficient operation of electrical power systems.

Overhead power transmission lines.   The transmission of electrical power requires many long interconnected power lines to carry the electrical current from where it is produced to where it is used, as shown in Figure 7-2. However, overhead power transmission lines require much planning to assure the best use of our land. The location of overhead transmission lines is limited by zoning laws and by populated areas, highways, railroads, and waterways, as well as other topographical and environmental factors. Today, an increased emphasis is being placed on environmental and aesthetic factors. Overhead power transmission lines ordinarily operate at voltage

Figure 7-2 Electrical power transmission lines. (Courtesy of Lapp Insulator, Interpace Corp.)

levels from 12 to 500 kV. Common transmission line voltages are in the range 50 to 150 kV.

Underground power transmission. Underground transmission methods for urban and suburban areas must be considered since the right-of-way for overhead transmission lines is limited. One advantage of overhead cables is their ability to dissipate heat. The use of underground cable is ordinarily confined to the short distance required in congested urban areas. The cost of underground cable is much more than for overhead cable. To improve underground cable power-handling capability, research is being done in forced-cooling techniques, such as circulating-oil and with compressed-gas insulation. Another possible method is the use of cryogenic cables or superconductors which operate at extremely low temperatures and have a large power-handling capability.

Parallel operation of power systems. Electrical power distribution systems are operated in a parallel circuit arrangement. By adding more power sources (generators) in parallel, a greater load demand or current requirement can be met. On a smaller scale, this is like connecting two or more batteries in parallel to provide greater current capacity. Two parallel-connected three-phase alternators are shown schematically in Figure 7-3. Most power plants usually have two or more alternators connected to a set of power lines inside the plant. These power lines or "bus" lines are usually large copper bar conductors which can carry very high amounts of current. At low-load demand times, only one alternator would be connected to the bus lines.

Figure 7-4 expands the concept of parallel-connected systems. A pictorial illustration of two power plants joined together through a distribution substation is shown. The two power plants might be located 100 miles apart, yet they are connected

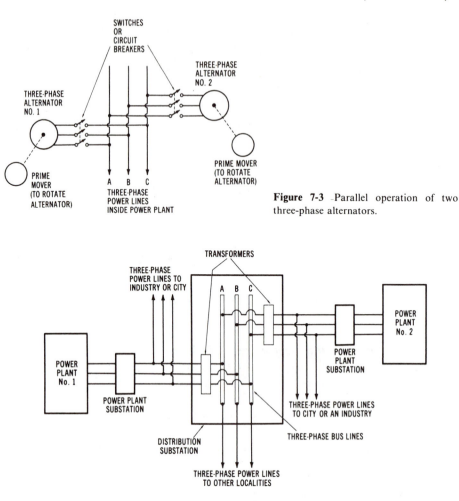

Figure 7-3 Parallel operation of two three-phase alternators.

Figure 7-4 Parallel operation of two electrical power plants.

in parallel to supply power to a specified region. If, for some reason (such as repairs on an alternator), the output of one power plant is reduced, the other power plant is still available to supply power to other localities. It is also possible for power plant 1 to supply part of the load requirement ordinarily supplied by power plant 2, or vice versa. These regional distribution systems of parallel-connected power sources provide automatic compensation for increased load demand in any area.

The major problem of parallel-connected distribution systems occurs when excessive load demands are encountered by several power systems in a single region. If all of the power plants in one area are operating near their peak power-output capacity, there is no backup capability. The equipment-protection system for each power plant and also for each alternator in the power plant is designed to disconnect from the system when its maximum power limits are reached. When the power demand on one part of the system becomes too excessive, the protective equipment will disconnect that part of the system. This places an even greater load on the remaining parts of the system. The excessive load now could cause other parts of the system to disconnect. This cycle could continue until the entire system is inoperative. This is what occurs when "blackouts" take place. No electrical power can be supplied to any part of the system until most of the power plants are put back in operation. The process of putting the output of a power plant back "on-line" when the system is down during power outages can be a long and difficult procedure.

TYPES OF DISTRIBUTION SYSTEMS

There are three general classifications of electrical power distribution systems. These are the radial, ring, and network systems shown in Figures 7-5 through 7-7. *Radial* systems are the simplest type since the power comes from one power source. A generating system supplies power from the substation through radial lines which are

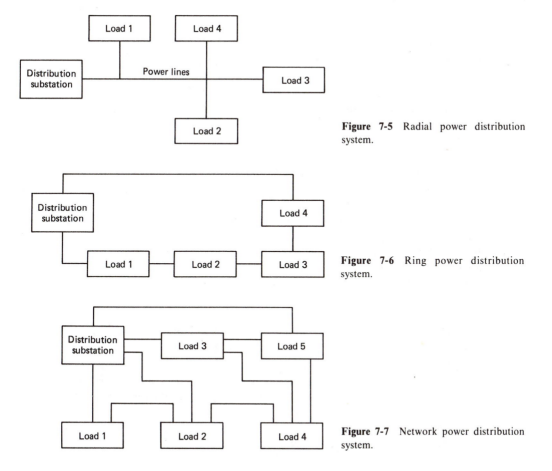

Figure 7-5 Radial power distribution system.

Figure 7-6 Ring power distribution system.

Figure 7-7 Network power distribution system.

extended to the various areas of a community (Figure 7-5). Radial systems are the least reliable in terms of continuous service since there is no backup distribution system. If any power line opens, one or more loads are interrupted. There is more likelihood of power outages. However, the radial system is the least expensive. This system is used in remote areas where other distribution systems are not economically feasible.

Ring distribution systems (Figure 7-6) are used in heavily populated areas. The distribution lines encircle the service area. Power is delivered from one or more power sources into substations near the service area. The power is then distributed from the substations through the radial power lines. When a power line is opened, no interruption of other loads occurs. The ring system provides a more continuous service than the radial system. Due to additional power lines and a greater circuit complexity, the ring system is more expensive.

Network distribution systems (Figure 7-7) are a combination of the radial and ring systems. They usually result when one of the other systems is expanded. Most of the distribution systems in the United States are network systems. This type of system is more complex but it provides very reliable service to consumers. With a network system, each load is fed by two or more circuits.

SPECIALIZED POWER DISTRIBUTION EQUIPMENT

To distribute electrical power, many types of specialized equipment are required. The electrical power system consists of such specialized equipment as overhead transmission lines, underground cables, transformers, fuses, circuit breakers, lightning arresters, power-factor correcting capacitors, and power metering systems.

Substations.   Substations contain many types of specialized power distribution equipment. They are an important part of electrical power distribution systems. The link between the high-voltage overhead transmission lines and low-voltage distribution is provided by substations. The function of a substation is to receive power from a high-voltage transmission system and convert it to voltage levels suitable for industrial, commercial, or residential distribution. The major functional component of a substation is the power transformer, which was discussed in Chapter 6. However, many other specialized types of equipment are required at a substation. A unit substation is shown in Figure 7-8.

High-voltage fuses.   Since short-circuited power lines are sometimes encountered, various protective devices are used to prevent damage to power lines and equipment. This protective equipment must be designed to handle high voltages and currents. Either fuses or circuit breakers may be used to protect high-voltage power lines. High-voltage fuses (those used for over 600 V) are made in several ways. An expulsion-type fuse has an element that will melt and vaporize when it is overloaded, causing the power line it is connected in series with to open. Liquid fuses are used to immerse a melted fuse element in a liquid to extinguish the arc developed by a faulty

Figure 7-8 Unit substation. (Courtesy of Square-D Co.)

current in the power line. A solid fuse is similar to a liquid fuse except that the arc is extinguished in a chamber filled with solid material. Some high-voltage fuses are shown in Figure 7-9.

Ordinarily, high-voltage fuses at outdoor substations are mounted adjacent to air-break disconnect switches. These switches provide a means of switching lines and disconnecting lines for repair. The fuse and switch enclosures are usually mounted near the overhead power lines at a substation.

Figure 7-9 High-voltage fuses. (Courtesy of Square-D Co.)

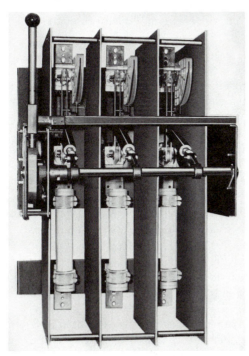

High-voltage circuit breakers.   Circuit breakers are also located at electrical substations. Many outdoor substations use oil-filled circuit breakers. In this type of circuit breaker, the contacts are immersed in an insulating oil contained in the metal enclosure. Another type of high-voltage circuit breaker is the magnetic blowout coil, which develops a magnetic air breaker in which the contacts separate in air when the power line is overloaded. The arc that results when the contacts open is extinguished by magnetic blowout coils, which develop a magnetic field that concentrates the arc into arc chutes. A modification of this type is the compressed-air circuit breaker. In this type, a stream of compressed air is concentrated on the contacts when the power line is opened. The compressed air aids in extinguishing the arc that is developed when the contacts open. It should be pointed out that large arcs are present whenever a high-voltage circuit is interrupted. This problem is not encountered to any great extent in low-voltage protective equipment.

Lightning arresters.   Lightning arresters are also part of outdoor substations. Lightning is a major cause of short-circuit situations on overhead power lines. The purpose of lightning arresters is to provide a path to ground for flashes of lightning. This eliminates flashover between the power lines, which causes short circuits. Valve-type lightning arresters are used frequently. They are two-terminal devices in which one terminal is connected to the power line and the other is connected to ground. The path from line to ground is of such high resistance that it is normally open. However, when lightning, which is a very high voltage, strikes a power line, it causes conduction from line to ground before flashover between lines occurs. After the lightning surge has been conducted to ground, the valve assembly then causes the arrester to be non-conductive once more.

Insulators and conductors.   All power lines must be isolated so as not to be safety hazards. Large strings of insulators are used at substations and at other points along the power distribution system to isolate the current-carrying conductors from their steel supports or any other ground-mounted equipment. Insulators may be made of porcelain, rubber, or thermoplastic material.

The conductors used for power distribution are ordinarily uninsulated aluminum or aluminum-conductor steel-reinforced (ACSR) for long-distance transmission, and insulated copper for shorter distances.

Distribution switching equipment.   Power distribution systems must have equipment for connecting or disconnecting the entire system or parts of the system. Various types of switching devices are used to perform this function. A simple type of switch is the safety switch that is mounted in a metal enclosure and operated by means of an external handle. The exterior and interior of a safety switch is shown in Figure 7-10. Safety switches are used only to turn a circuit off or on; however, fuses are often mounted in the same enclosure.

Another type of switching equipment is the kind used in conjunction with a circuit-breaker panelboard. Panelboards are metal cabinets enclosing a main disconnect switch and branch-circuit protective equipment. Distribution panelboards

(a)   **(b)**

Figure 7-10 Safety switch: (a) exterior; (b) interior. (Courtesy of Square-D Co.)

Figure 7-11 Power distribution load center. (Courtesy of Square-D Co.)

are usually located between the power feed lines within a building and branch circuits that supply the equipment in a section of the building. Circuit breakers are placed inside the panelboard to protect each of the branch circuits connected.

Metal-enclosed, low-voltage switchgear is used in many industrial and commercial buildings. Such switchgear is used as a distribution control center to house the circuit breakers, bus bars, and terminal connections that are part of the industrial distribution systems. Ordinarily, a combination of switchgear and distribution transformers is placed in adjacent metal enclosures, such as that shown in Figure 7-11. Such a combination is referred to as a *load center unit substation* since it is the central control for several loads. The rating of these load centers is usually 15,000 V or less for the high-voltage section, and 600 V or less for the low-voltage section. Load centers provide flexibility in the electrical power distribution design of industrial and commercial buildings.

DISTRIBUTION IN INDUSTRIAL AND COMMERCIAL BUILDINGS

Electrical power is delivered to industries and commercial buildings and then distributed by the power system within the building. Various types of circuit breakers and swtichgear are employed for interior distribution. Another factor involved in industrial and commercial power distribution is the distribution of electrical energy to the many types of loads connected to the system. This part of the distribution system is concerned with the conductors, feeder systems, branch circuits, grounding methods, and machine protective and control devices used.

Raceways.   Much electrical distribution in large industrial and commercial buildings is contained in *raceways*. Raceways may be large metal ducts, such as the one shown in Figure 7-12, or rigid metal conduit. Raceways contain the conductors that distribute power to the various equipment within a building. Copper conductors are ordinarily used for interior power distribution. The physical size of each conductor required is dependent on the current rating of the branch circuit.

Figure 7-12 Raceway. (Courtesy of Square-D Co.)

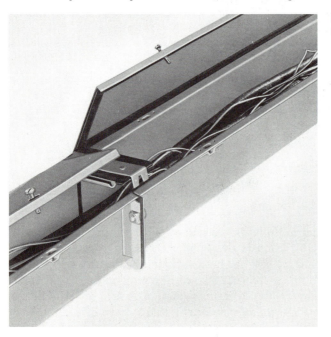

Feeders and branch circuits.   The conductors that carry current to the electrical load devices in buildings are called *feeders* and *branch circuits.* Feeder lines supply power to branches that are connected to them. Primary feed lines in buildings may be either overhead or underground. In several situations, overhead lines are preferred because they add flexibility for future expansion. Underground systems are more expensive, but present a better appearance. Secondary feeders are connected to the primary feeder lines to supply power to individual sections within the building. Either aluminum or copper feeder lines may be used, depending on the specific power requirements. The distribution from the feeder lines is then connected through individual protective equipment to branch circuits that supply the various industrial loads. Each branch circuit has protective devices according to the needs of that particular branch. The overall feeder-branch system in a building may be a very complex network of switching equipment, transformers, conductors, and protective equipment.

DISTRIBUTION SYSTEM GROUNDING

An important factor in the operation of any electrical distribution system is grounding. Proper grounding techniques are required for safety as well as circuit performance. There are two types of grounding: (1) system grounding and (2) equipment grounding. Another important grounding consideration is ground-fault protection.

System grounding.   System grounding involves the actual grounding of a current-carrying conductor (usually called the neutral) of a power system. Three-phase systems may be either the wye or delta type. The wye system has an obvious advantage over the delta system since one side of each phase winding is connected to ground. A ground is defined as a reference point of zero volts potential, which is usually an actual connection to earth ground. The common terminals of the wye system, when connected to ground, become the neutral conductor of the three-phase four-wire system.

The delta system does not readily lend itself to grounding since it does not have a common neutral. The problem of ground faults (line-to-ground shorts) occurring in underground delta systems is much greater than in wye systems. A common method of grounding a delta system is to use a wye-delta transformer connection and ground the common terminals of the wye-connected primary. However, the wye system is now used more often for industrial distribution, since the secondary is easily grounded and it provides overvoltage protection from lightning or line-to-ground shorts.

Single-phase 120/240- or 240/480-V systems are grounded in a manner similar to a three-phase ground. The neutral of the single-phase three-wire system is grounded by a metal rod driven into the earth at the transformer location. System grounding conductors are insulated with white material for easy identification.

Equipment grounding.   The second type of ground is the equipment ground which, as the term implies, places operating equipment at ground potential. The conductor used for this purpose is either bare wire or a green insulated wire. The *National Electrical Code®* (NEC) describes conditions that require fixed electrical equipment to be grounded. Usually, all fixed electrical equipment located in industrial plants should be grounded. Types of equipment that should be grounded include enclosures for switching and protective equipment for load control, transformer enclosures, electric motor frames, and fixed electronic test equipment. All buildings should use 120-V single-phase duplex receptacles of the grounded type for all portable tools. The proper grounding of these receptacles can be checked by using a plug-in tester.

Ground-fault protection.   Ground-fault interrupters (GFIs) such as the one shown in Figure 7-13 are now used extensively in power distribution systems. It is required by the NEC that all 120-V single-phase 15- or 20-A receptacle outlets that are outdoors or in bathrooms have ground-fault interrupters installed. These devices are designed to reduce electrical shock hazards resulting from people coming in contact with a hot ac line (line-to-ground short). The circuit interrupter is designed to sense any change in circuit conditions, such as would occur when a line-to-ground short exists.

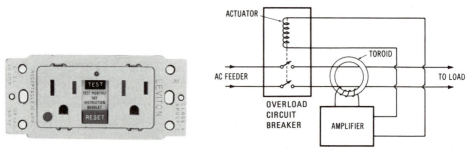

Figure 7-13 Ground-fault interrupter. (Courtesy of Leviton Co.)

Figure 7-14 Schematic of a ground-fault interrupter.

One type of GFI has control wires that extend through a magnetic toroidal loop (see Figure 7-14). Ordinarily, the ac current flowing through the conductors inside the loop is equal in magnitude and opposite in direction. Any change in this equal and opposite condition is sensed by the magnetic toroidal loop. When a line-to-ground short occurs, an instantaneous change in circuit conditions occurs. The change causes a magnetic field to be induced into the toroidal loop. The induced current is amplified to a level sufficient to cause the circuit breaker to open. Thus any line-to-ground short will cause the ground-fault interrupter to open. The operating speed is so fast (only a very small current opens the circuit) that the shock hazard is greatly reduced.

Construction sites, where temporary wiring is set up, are also required to use GFIs for protection of workers using electrical equipment. Ground-fault protection of industrial equipment is provided for wye-connected systems of 150 to 600 V for each distribution panelboard rated over 1000 A. In this situation, the GFI will open all ungrounded conductors at the panelboard when a line-to-ground short occurs.

DISTRIBUTION SYSTEM WIRING DESIGN

The wiring design of electrical power distribution systems can be very complex. There are many factors that must be considered in the wiring design of a distribution system installed in a building. Wiring design standards are specified in the *National Electrical Code®*, which is published by the National Fire Protection Association (NFPA). The NEC, local wiring standards, and electrical inspection policies should be considered in distribution system wiring design.

There are several distribution system wiring design considerations which are pointed out specifically in the NEC. In this section, voltage-drop calculations, branch-circuit design, feeder-circuit design, and the design for grounding systems are discussed.

The National Electrical Code®. The NEC sets forth the minimum standards for electrical wiring in the United States. The standards contained in the NEC are enforced by being incorporated into the various city and community ordinances which deal with electrical wiring in residences, industrial plants, and commercial buildings. Therefore, these local ordinances conform to the standards set forth in the NEC.

In most areas of the United States, a license must be obtained by any person who does electrical wiring. Usually, a test administered by the city, county, or state must be passed in order to obtain this license. These tests are based on local ordinances and the NEC. The rules for electrical wiring that are established by the local electrical power company are also sometimes incorporated into the license test.

Electrical inspections. When new buildings are constructed, they must be inspected to see if the electrical wiring meets the standards of the local ordinances, the NEC, and the local power company. The organization that supplies the electrical inspectors varies from one locality to another. Ordinarily, the local power company can advise people about whom to contact for information about electrical inspections.

Voltage Drop in Electrical Distribution Systems

Although the resistance of electrical conductors is very low, a long length of wire could cause a substantial voltage drop. This is illustrated in Figure 7-15. Remember that a voltage drop is current times resistance ($I \times R$). Therefore, whenever current flows through a circuit, a voltage drop is created. Ideally, the voltage drop caused by the resistance of a conductor will be very small.

However, a longer section of electrical conductor has a higher resistance.

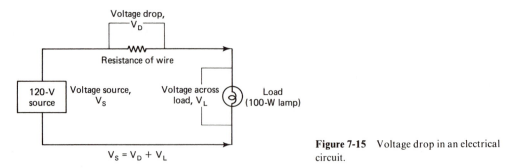

Figure 7-15 Voltage drop in an electrical circuit.

Therefore, it is sometimes necessary to limit the distance a conductor can extend from the power source to the load that it supplies. Many types of loads do not operate properly when a value less than the full source voltage is available.

Observe from Figure 7-15 that as the voltage drop (V_D) increases, the voltage applied to the load (V_L) decreases. As current in the circuit increases, V_D increases causing V_L to decrease since the source voltage remains constant.

Voltage-drop calculations using a conductor table.   It is important when dealing with electrical wiring design to be able to determine the amount of voltage drop caused by conductor resistance. Table 3-2 may be used to make these calculations. The NEC limits the amount of voltage drop that a distribution system can have. This means that long runs of conductors must ordinarily be avoided. Remember that a conductor with a large cross-sectional area will cause a smaller voltage drop since its resistance is smaller.

To better understand how to determine the size of conductor required to limit the voltage drop in a system, a sample problem is illustrated below.

Given: A 200-A load located 400 ft from a 240-V single-phase source. Limit the voltage drop to 2% of the source voltage.
Find: The size of copper conductors with RH insulation needed to limit the voltage drop of the distribution system.
Solution:

1. The allowable voltage drop equals 240 V times 0.02 (2%). This equals 4.8 V.
2. Determine the maximum resistance for 800 ft. This is the equivalent of 400 ft $\times$ 2, since there are two current-carrying conductors for a single-phase system.

$$R = \frac{V_D}{I}$$

$$= \frac{4.8 \text{ V}}{200 \text{ A}}$$

$$= 0.024 \ \Omega \text{ (resistance for 800 ft)}$$

3. Determine the maximum resistance for 1000 ft of conductor.

$$\frac{800 \text{ ft}}{1000 \text{ ft}} = \frac{0.024 \text{ } \Omega}{R}$$

$$800 R = (1000)(0.024)$$

$$R = 0.030 \text{ } \Omega$$

4. Use Table 3-2 to find the size of copper conductor which has the nearest dc resistance (Ω/1000 ft) value that is *equal to or less than* the value calculated in step 3. The conductor chosen is conductor size 350 MCM, RH copper.

5. Check this conductor size in an ampacity table to assure that it is large enough to carry 200 A. Table 3-3 shows that a 350-MCM RH copper conductor will handle 310 A of current; therefore, use 350-MCM conductors.
   (Always remember to use the largest conductor when steps 4 and 5 produce different values.)

Alternative method of voltage-drop calculation. In some cases, an easier method to determine the conductor size for limiting the voltage drop is to use one of the following formulas to find the cross-sectional (cmil) area of the conductor:

$$\text{cmil} = \frac{p \times I \times 2d}{V_D} \qquad \text{(single-phase systems)}$$

or

$$\text{cmil} = \frac{p \times I \times 1.73d}{V_D} \qquad \text{(three-phase systems)}$$

where p = resistivity from Table 3-1
I = load current in amperes
V_D = allowable voltage drop
d = distance from source to load in feet

The sample problem given for a single-phase system in the preceding section could be set up as follows:

$$\text{cmil} = \frac{p \times I \times 2d}{V_D}$$

$$= \frac{10.4 \times 200 \times 2 \times 400}{240 \times 0.02}$$

$$= \frac{1,664,000}{4.8}$$

$$= 346,666$$

$$= 347 \text{ MCM}$$

The next largest size is a 350-MCM conductor.

Branch-Circuit Distribution Design Considerations

A branch circuit is defined as a circuit that extends from the last overcurrent protective device of the power distribution system. Branch circuits, according to the NEC, are either 15, 20, 30, 40, or 50 A in capacity. Loads larger than 50 A are not connected to a branch circuit.

There are many rules in the NEC which apply to branch-circuit design. The following information is based on the NEC. First, each circuit must be designed so that accidental short circuits or grounds do not cause damage to any part of the system. Then, fuses or circuit breakers are to be used as branch-circuit overcurrent protective devices. Should a short circuit or ground condition occur, the protective device should open and interrupt the flow of current in the branch circuit. One important NEC rule is that No. 16 or No. 18 (extension cord) wire may be extended from No. 12 or No. 14 conductors but not larger than No. 12. This means that an extension cord of No. 16 wire should not be plugged into a receptacle that uses No. 10 wire. Damage to smaller wires (due to the heating effect) before the overcurrent device can open is eliminated by applying this rule. Lighting circuits are one of the most common types of branch circuits. They are usually either 15- or 20-A circuits.

The maximum rating of an individual load (such as a portable appliance connected to a branch circuit) is 80% of the branch circuit current rating. So a 20-A circuit could not have a single load that draws more than 16 A. If the load is a permanently connected appliance, its current rating cannot be more than 50% of the branch circuit capacity if portable appliances or lights are connected to the same circuit.

Voltage drop in a branch circuit.   Branch circuits must be designed so that sufficient voltage is supplied to all parts of the circuit. The distance that a branch circuit can extend from the voltage source or power distribution panel is therefore limited. A voltage drop of 3% is specified by the NEC as the maximum allowed for branch circuits in electrical wiring design.

Figure 7-16 Voltage-drop calculation for a branch circuit.

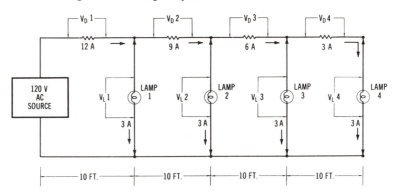

The method for calculating the voltage drop in a branch circuit is a step-by-step process which is illustrated by the following problem. Refer to the circuit diagram given in Figure 7-16.

Given: A 120-V 15-A branch circuit supplies a load which consists of four lamps. Each lamp draws 3 A of current from the source. The lamps are located at 10-ft intervals from the power distribution panel.
Find: The voltage across lamp 4.
Solution:

1. Find the resistance for 20 ft of conductor (same as 10-ft conductor × 2). A No. 14 copper wire is used for 15-A branch circuits. From Table 3-2, find that the resistance of 1000 ft of No. 14 copper wire is 2.57 Ω. Therefore, the resistance of 20 ft of wire is

$$\frac{20 \text{ ft}}{1000 \text{ ft}} = \frac{R}{2.57 \text{ } \Omega}$$
$$1000\,R = (20)(2.57)$$
$$R = 0.0514 \text{ } \Omega$$

2. Calculate voltage drop $V_D\,1$ (R equals the resistance of 20 ft of wire).

$$V_D\,1 = I \times R$$
$$= 12\text{A} \times 0.0514$$
$$= 0.6168 \text{ V ac}$$

3. Calculate load voltage $V_L\,1$ (the source voltage minus $V_D\,1$).

$$V_L\,1 = 120 \text{ V} - 0.6168 \text{ V}$$
$$= 119.383 \text{ V}$$

4. Calculate voltage drop $V_D\,2$ and load voltage $V_L\,2$.

$$V_D\,2 = I \times R$$
$$= 9\text{A} \times 0.0514$$
$$= 0.4626 \text{ V ac}$$
$$V_L\,2 = 119.383 \text{ V} - 0.4626 \text{ V}$$
$$= 118.920 \text{ V}$$

5. Calculate voltage drop $V_D\,3$ and load voltage $V_L\,3$.

$$V_D\,3 = I \times R$$
$$= 6\text{A} \times 0.0514$$
$$= 0.3084 \text{ V ac}$$
$$V_L\,3 = 118.920 \text{ V} - 0.3084 \text{ V}$$
$$= 118.612 \text{ V}$$

6. Calculate voltage drop $V_D\,4$ and load voltage $V_L\,4$.

$$V_D\,4 = I \times R$$
$$= 3\text{A} \times 0.0514$$
$$= 0.1542 \text{ V ac}$$
$$V_L\,4 = 118.612 \text{ V} - 0.1542 \text{ V}$$
$$= 118.458 \text{ V}$$

Notice that the voltage across lamp 4 is substantially reduced from the 120-V source value due to the voltage drop of the conductors. Also, notice that the resistances used to calculate the voltage drops represented both wires (hot and neutral) of the branch circuit. Ordinarily, 120-V branch circuits do not extend more than 100 ft from the power distribution panel. The preferred distance is 75 ft. The voltage drop in branch-circuit conductors can be reduced by making the circuit shorter in length or by using larger conductors.

In residential electrical wiring distribution design, the voltage drop in many branch circuits is difficult to calculate since the lighting and portable appliance receptacles are placed on the same branch circuits. Since portable appliances and "plug-in" lights are not used all of the time, the voltage drop will vary according to the number of lights and appliances in use. This problem is usually not encountered in an industrial or commercial wiring design for lights, since the lighting units are usually larger and are permanently installed on the branch circuits.

Branch-circuit wiring. A branch circuit usually consists of nonmetallic-sheathed cable which is connected into a power distribution panel. Each branch circuit that is wired from the power distribution panel is protected by a fuse or circuit breaker.

Figure 7-17 Single-phase three-wire power distribution panel.

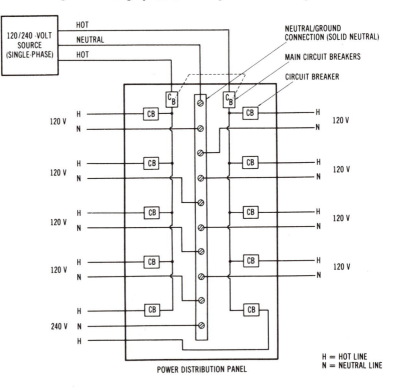

The power panel also has a main switch which controls all of the branch circuits that are connected to it.

Single-phase branch circuits.   A diagram of a single-phase three-wire (120/240-V) power distribution panel is shown in Figure 7-17. Notice that eight 120-V branch circuits and one 240-V circuit are available from the power panel. This type of system is used in most homes where several 120-V branch circuits and, typically, three or four 240-V branch circuits are required. Notice in Figure 7-17 that each hot line has a circuit breaker while the neutral line connects directly to the branch circuits. Neutrals should never be opened (fused). This is a safety precaution in electrical wiring design.

Three-phase branch circuits.   A diagram for a three-phase four-wire (120/208 V) power distribution panel is shown in Figure 7-18. There are three single-phase 120-V branch circuits and two three-phase 208-V branch circuits shown. The single-phase branches are balanced (one hot line from each branch). Each hot line has an individual circuit breaker. Three-phase lines should be connected so that an overload in the branch circuit will cause all three-phase lines to open. This is accomplished by using a three-phase circuit breaker which is arranged internally as shown in Figure 7-19.

Figure 7-18   Three-phase four-wire power distribution panel.

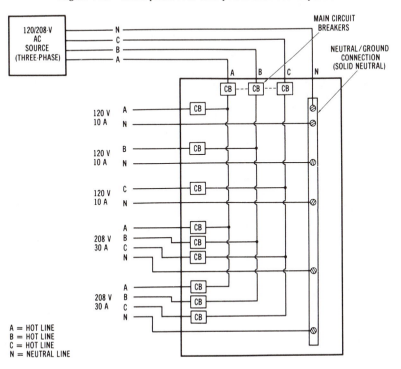

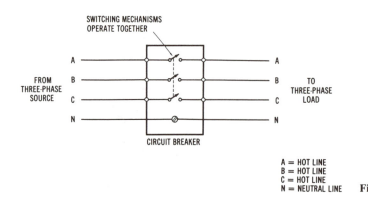

SWITCHING MECHANISMS
OPERATE TOGETHER

FROM
THREE-PHASE
SOURCE

TO
THREE-PHASE
LOAD

CIRCUIT BREAKER

A = HOT LINE
B = HOT LINE
C = HOT LINE
N = NEUTRAL LINE

Figure 7-19 Three-phase circuit breaker.

Feeder-Circuit Design Considerations

Feeder circuits are used to distribute electrical power to power distribution panels. Many feeder circuits extend for very long distances; therefore, voltage drop must be considered in feeder-circuit design. In high-voltage feeder circuits, the voltage drop is reduced. However, many lower-voltage feeder circuits require large-diameter conductors to provide a tolerable level of voltage drop. High-current feeder circuits also present a problem in terms of the massive overload protection that is sometimes required. This protection is usually provided by system switchgear or load center where the feeder circuits originate.

Determining the size of feeder circuits. The amount of current that a feeder circuit must be designed to carry depends on the actual load demanded by the branch-circuit power distribution panels which it supplies. Each power distribution panel will have a separate feeder circuit. Also, each feeder circuit must have its own overload protection.

The following problem is an example of conductor size calculation for a feeder circuit.

Given: Three 15-kW fluorescent lighting units are connected to a three-phase four-wire (277/480-V) system. The lighting units have a power factor of 0.8.
Find: The size of aluminum feeder conductors with THW insulation required to supply this load.
Solution:

1. Find the line current:

$$I_L = \frac{P_T}{1.73 \times V_L \times PF}$$

$$= \frac{45,000 \text{ W}}{1.73 \times 480 \text{ V} \times 0.8}$$

$$= 67.74 \text{ A}$$

2. From Table 3-3, find that the conductor size which will carry 67.74 A of current is a No. 3 AWG THW aluminum conductor.

Voltage-drop calculation for a feeder circuit. Feeder circuit design must take the conductor voltage drop into consideration. The voltage drop in a feeder circuit must be kept as low as possible so that maximum power can be delivered to the loads connected to the feeder system. The NEC allows a maximum 5% voltage drop in the combination of a branch and a feeder circuit; however, a 5% voltage reduction represents a significant power loss in a circuit. Power loss due to voltage drop can be calculated as V^2/R, where V is the voltage drop of the circuit and R is the resistance of the conductors of the circuit.

The calculation of feeder conductor size is similar to that for a branch circuit voltage drop. The size of the conductors must be large enough to (1) have the required ampacity and (2) keep the voltage drop below a specified level. If the second requirement is not met, possibly due to a long feeder circuit, the conductors chosen must be larger than the required ampacity rating.

The following problem illustrates the calculation of feeder conductor size based on the voltage drop in a single-phase circuit.

Given: A single-phase 240-V load in a factory is rated at 85 kW. The feeders (two hot lines) will be 260-ft lengths of RHW copper conductor. The maximum conductor voltage drop allowed is 2%.
Find: The feeder conductor size required.
Solution:

1. Find the maximum voltage drop of the circuit.
$$V_D = \% \times \text{load}$$
$$= 0.02 \times 240$$
$$= 4.8 \text{ V}$$

2. Find the current drawn by the load.
$$I = \frac{\text{power}}{\text{voltage}}$$
$$= \frac{85,000 \text{ W}}{240 \text{ V}}$$
$$= 354.2 \text{ A}$$

3. Find the minimum circular-mil conductor area required. Use the formula given for finding the cross-sectional area of a conductor in single-phase systems.
$$\text{cmil} = \frac{p \times I \times 2d}{V_D}$$
$$= \frac{10.4 \times 354.2 \times 2 \times 260}{4.8}$$
$$= 399,065.33 \text{ cmil}$$

4. Determine the feeder conductor size. The next larger size conductor in Table 3-2 is 400 MCM. Check Table 3-3 and see that a 400-MCM RHW copper conductor will carry 335 A. This is less than the required 354.2 A, so use the next larger size, which is a 500-MCM conductor.

The conductor size for a three-phase feeder circuit is determined in a similar way. In this problem, the feeder size will be determined as based on the circuit voltage drop.

Given: A 480-V three-phase three-wire (delta) feeder circuit supplies a 45-kW balanced load to a commercial building. The feeder circuit (three hot lines) will be a 300-ft length of RH copper conductor. The maximum voltage drop is 1%.
Find: The feeder size required (based on the voltage drop of the circuit).
Solution:

1. Find the maximum voltage drop of the circuit.
$$V_D = 0.01 \times 480$$
$$= 4.8 \text{ V}$$

2. Find the line current drawn by the load.
$$I_L = \frac{P}{1.73 \times V \times \text{PF}}$$
$$= \frac{45,000}{1.73 \times 480 \times 0.75}$$
$$= 72.25 \text{ A}$$

3. Find the minimum circular-mil conductor area required. Use the formula for finding cmil in three-phase systems.
$$\text{cmil} = \frac{p \times I \times 1.73d}{V_D}$$
$$= \frac{10.4 \times 72.25 \times 1.73 \times 300}{4.8}$$
$$= 8125 \text{ cmil}$$

4. Determine the feeder conductor size. The closest and next larger conductor size in Table 3-2 is No. 1 AWG. Check Table 3-3 and see that a No. 1 AWG RH copper conductor will carry 130 A. This is much more than the required 72.25 A. Therefore, use No. 1 AWG RH copper conductors for the feeder circuit.

Determining Grounding-Conductor Size

Grounding considerations in electrical wiring design were discussed previously. Another aspect in wiring design is to determine the size of the grounding conductor required in a circuit. All circuits that operate at 150 V or less must be grounded; therefore, all residential electrical systems must be grounded. Higher-voltage systems used in industrial and commercial buildings have grounding requirements that are specified by the NEC and by local codes. A ground at the service entrance of a building is usually a metal water pipe which extends uninterrupted underground or a grounding electrode that is driven into the ground near the electrical service entrance.

The size of the grounding conductor is determined by the current rating of the distribution system. Table 7-1 lists equipment grounding-conductor size for interior wiring and Table 7-2 lists the minimum grounding-conductor sizes for system

TABLE 7-1 EQUIPMENT GROUNDING
CONDUCTOR SIZES FOR INTERIOR WIRING

Ampere Rating of Distribution Panel	Grounding Conductor Size (AWG)*	
	Copper	Aluminum
15	14	12
20	12	10
30	10	8
40	10	8
60	10	8
100	8	6
200	6	4
400	3	1
600	1	00
800	0	000
1000	00	0000

* No. 12 or No. 14 wiring cable can use a No. 18 equipment ground.

TABLE 7-2 SYSTEM GROUNDING-CONDUCTOR
SIZES FOR SERVICE ENTRANCES

Conductor Size (Hot Line)		Grounding Conductor Size	
Copper	Aluminum	Copper (AWG)	Aluminum (AWG)
No. 2 AWG or smaller	No. 0 AWG or smaller	8	6
No. 1 or 0 AWG	No. 00 or 000 AWG	6	4
No. 00 or 000 AWG	No. 0000 AWG or 250 MCM	4	2
No. 000 AWG to 350 MCM	250 MCM to 500 MCM	2	0
350 MCM to 600 MCM	500 MCM to 900 MCM	0	000
600 MCM to 1100 MCM	900 MCM to 1750 MCM	00	0000

grounding of service entrances. The sizes of grounding conductors listed in Table 7-1 are for equipment grounds that connect to raceways, enclosures, and metal frames for safety purposes. Note that a No. 12 or a No. 14 wiring cable, such as the type commonly used for house wiring, can have a No. 18 equipment ground. The ground is contained in the same cable as the hot conductors. Table 7-2 is used to find the minimum size of grounding conductors needed for service entrances, based on the size of the hot line conductors used with the system.

PARTS OF INTERIOR ELECTRICAL DISTRIBUTION SYSTEMS

Some parts of interior electrical distribution systems have been discussed previously. Such types of equipment as transformers, switchgear, conductors, insulators, and protective equipment are parts of interior distribution systems. There are, however, certain parts of interior electrical distribution systems which are unique to the wiring system itself. These parts include the nonmetallic-sheathed cable (NMC), the metal-clad cable, the rigid conduit, and the electrical metallic tubing (EMT).

Nonmetallic-sheathed cable (NMC). Nonmetallic-sheathed cable is a common type of electrical cable used for interior wiring. NMC, sometimes referred to as *Romex* cable, is used in residential wiring systems almost exclusively. The most common type used is No. 12-2 WG, which is illustrated in Figure 7-20. This type of NMC comes in 250-ft rolls for interior wiring. The cable has a thin plastic outer covering with three conductors inside. The conductors have colored insulation which designates whether the conductor should be used as a hot, neutral, or equipment ground wire. For instance, the conductor connected to the hot side of the system has black or red insulation, while the neutral conductor has white or gray insulation. The equipment grounding conductor has either green insulation or no insulation (bare conductor). There are several different sizes of bushings and connectors used for the installation of NMC in buildings.

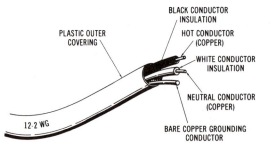

Figure 7-20   Nonmetallic sheathed cable.

The designation No. 12-2 WG means that (1) the copper conductors used are No. 12 AWG as measured by an American Wire Gage (AWG), (2) there are two current-carrying conductors, and (3) the cable comes with a ground (WG) wire. A No. 14-3 WG cable, in comparison, would have three No. 14 conductors and a grounding conductor. NMC ranges in size from No. 14 to No. 1 AWG copper conductors and from No. 12 to No. 2 AWG aluminum conductors.

Metal-clad cable.   Metal-clad cable is similar to NMC except that it has a flexible spiral metal covering rather than a plastic covering. A common type of metal-clad cable is called BX cable. Like NMC, BX cable contains two or three conductors. There are also several sizes of connectors and bushings used in the installation of BX cable. The primary advantage of this type of metal-clad cable is that it is contained in a metal enclosure that is flexible so that it can be bent easily. Other metal enclosures are usually more difficult to bend.

Rigid conduit.   The exterior of rigid conduit looks like water pipe. It is used in special locations for enclosing electrical conductors. Rigid conduit comes in 10-ft lengths which must be threaded for joining the pieces together. The conduit is secured to metal wiring boxes by locknuts and bushings. It is bulky to handle and takes a long time to install.

Electrical metallic tubing (EMT).   EMT or thin-wall conduit is somewhat like rigid conduit except that it can be bent with a special conduit-bending tool. EMT is easier to install than rigid conduit, since no threading is required. It also comes in 10-ft lengths. EMT is installed by using compression couplings to connect the conduit to metal wiring boxes. Interior electrical wiring systems use EMT extensively since it can be connected to metal wiring boxes.

REVIEW

7.1. What is the purpose of a power distribution system?

7.2. What are applications of single-phase power? Of three-phase power?

7.3. How are electrical power transmission systems interconnected?

7.4. Why is undergound power transmission used?

7.5. Discuss parallel operation of power systems.

7.6. Discuss the three general classifications of power distribution systems.

7.7. What is the purpose of an electrical distribution substation?

7.8. How do high-voltage fuses differ from low-voltage fuses?

7.9. What special design features do high-voltage circuit breakers have?

7.10. What is the purpose of a lightning arrestor?

7.11. What types of conductors are ordinarily used for long-distance power transmission?

7.12. What are some types of distribution switching equipment?

7.13. Define the following terms relating to electrical distribution systems: **(a)** raceway, **(b)** feeder circuit, **(c)** branch circuit, **(d)** system ground, and **(e)** equipment ground.

7.14. What is the purpose of ground-fault interrupters?

7.15. What is the *National Electrical Code®*?

7.16. Why is voltage drop important in electrical distribution systems?

7.17. How is the circular-mil area of a conductor of a three-phase distribution system calculated?

7.18. What are the typical ampacity ratings of branch circuits?

7.19. Discuss the following equipment which is used with interior electrical distribution systems: **(a)** NMC, **(b)** BX cable, **(c)** rigid conduit, and **(d)** EMT.

PROBLEMS

7.1. Find the proper size of THW copper conductor needed to limit to 2% the voltage drop of a 400-A load that is connected 150 ft from a 480-V single-phase source. Use Table 3-2 to find the connector size.

7.2. Three 20,000-W loads are connected in a system. The power factor of the system is 0.9. Find the minimum size of RH copper feeder conductors needed to supply this load.

7.3. Determine the branch-circuit rating and size of conductors needed for the following systems (based on 80% branch-circuit current requirement):
  (a) 2500-W 240-V heater   **(b)** 1800-W 120-V washer
  (c) 12,000-W 240-V range   **(d)** 800-W 120-V toaster

7.4. Calculate the maximum distance a 120-V 30-A branch circuit can extend from the power source when a 3-kW appliance is connected to the circuit. The voltage drop should be limited to 2%.

7.5. A 120-V 20-A branch circuit (using No. 12 copper conductors) extends 150 ft. Loads connected at 30-ft intervals each draw 3.5 A. With all loads connected, calculate the voltage at the last outlet.

7.6. Calculate the longest distance a single-phase 20-A 120-V branch circuit can extend from the power source. Use No. 12 AWG copper conductors and limit the voltage drop to 2%.

7.7. A single-phase 120-V load is rated at 40 kW. The feeder circuit will be 400 ft of RH copper conductors. Find the minimum conductor size necessary to supply this load and still limit the voltage drop to 1%.

7.8. A 240-V three-phase delta system converts 30 kW of power per phase (balanced). The system power factor is unit (1.0). The feeders will be 400 ft of RHW copper conductors. Find the minimum conductor size necessary to limit the voltage drop to 2%.

7.9. Find the minimum size of copper grounding conductor needed for the following situations.
  (a) No. 10 AWG copper service entrance   **(b)** 250-MCM aluminum service entrance
  (c) 200-A service entrance   **(d)** 500-MCM copper service entrance
  (e) 600-A service entrance

7.10. Find the sizes of copper equipment grounding conductor to be used for the following distribution panel current ratings.
  (a) 100 A
  (b) 200 A
  (c) 1000 A

EIGHT

Direct-Current Motors

Rotating machines that convert electrical energy into mechanical energy represent a broad category of electrical loads used today. The motor load is the major electrical power-consuming load used in the United States. Motors of various sizes are used for processes ranging from precise machinery control to the movement of massive pieces of equipment in industry. They are used in home appliances, commercial building heating and cooling systems, and for many other applications.

BASIC MOTOR PRINCIPLES

The function of a motor is to convert electrical energy into mechanical energy in the form of a rotary motion. To produce a rotary motion, a motor must have an electrical power input. Recall from Chapter 5 that generator action is brought about due to a magnetic field, a set of conductors within the magnetic field, and relative motion between the two. Motion is similarly produced in a motor due to the interaction of a magnetic field and a set of conductors. This interaction is referred to as *motor action*.

All motors, regardless of whether they operate from an ac or a dc power line, have several basic characteristics in common. Their basic parts include (1) a stator, which is the frame and other stationary components; (2) a rotor, which is the rotating shaft and its associated parts; and (3) auxiliary equipment, such as a brush/commutator assembly for dc motors and a starting circuit for single-phase ac motors. The basic parts of a dc motor are shown in Figure 8-1. A simple dc motor is constructed in the same way as a dc generator. Their basic parts are the same.

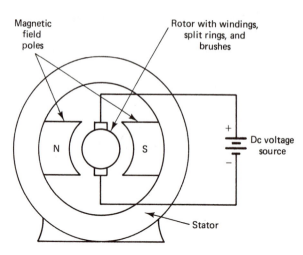

Magnetic
field
poles

Rotor with windings,
split rings, and
brushes

Dc voltage
source

Stator

Figure 8-1   Basic parts of a dc motor.

The motor principle (motor action) is illustrated in Figure 8-2. In Figure 8-2(a) no current is flowing through the conductors due to the position of the brushes in relation to the commutator. During this condition, no motion is produced. When current flows through the conductor, a circular magnetic field is developed around the conductor. The direction of the current flow determines the direction of the circular magnetic fields, as shown in the cross-sectional diagram of Figure 8-2(c).

When current flows through the conductors within the main magnetic field, this field interacts with the main field. The interaction of these two magnetic fields results in motion being produced. The circular magnetic field around the conductors causes a compression of the main magnetic flux at points A and B in Figure 8-2(b). This compression causes the magnetic field to produce a reaction in the opposite direction of the compression. Therefore, motion is produced away from points A and B. In actual motor operation, a rotary motion in a clockwise direction would be produced. The right-hand motor rule shown in Figure 8-2(d) is used to predict the direction of motion. To change the direction of rotation, the direction of the current flow through the conductors would have to be reversed.

The rotating effect produced by the interaction of two magnetic fields is called *torque* or *motor action*. The torque produced by a motor depends on the strength of the main magnetic field and the amount of current flowing through the conductors. As the magnetic field strength or the current through the conductors increases, the amount of torque or rotary motion will increase also.

TORQUE DEVELOPMENT IN DC MOTORS

Motor action or the development of torque is a significant electromechanical principle. There are some basic factors of torque development which should be kept in mind. Motor action allows electrical current to produce rotation (torque). During this process, a generated voltage or counterelectromotive force (CEMF) is produced. This

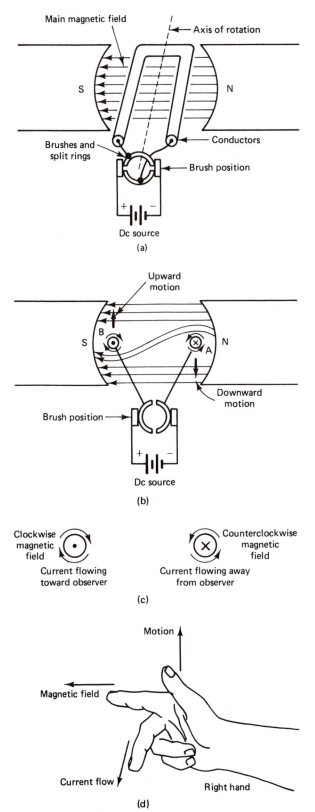

Main magnetic field

Axis of rotation

S

N

Brushes and split rings

Conductors

Brush position

Dc source

(a)

Upward motion

B

S

N

A

Brush position

Downward motion

Dc source

(b)

Clockwise magnetic field

Current flowing toward observer

Counterclockwise magnetic field

Current flowing away from observer

(c)

Motion

Magnetic field

Current flow

Right hand

(d)

Figure 8-2 Motor principle (motor action): (a) condition with no current flowing through conductors; (b) condition with current flowing through conductors; (c) direction of current flow through conductors determines direction of magnetic field around conductors; (d) right-hand motor rule.

phenomenon is called "generator action" of a motor. The generated voltage due to CEMF (V_c) is always less than the voltage applied to the motor (V_t) and its direction is such that it opposes the armature current (I_a) flow.

A comparison should be made between motor action and generator action. When a rotating electrical machine is operated as a dc generator, a torque is developed in the armature conductors (countertorque) which opposes the torque of the prime mover. The armature current of a dc generator flows in the same direction as the polarity of the generated voltage (V_g) and V_g *exceeds* the voltage applied to the load circuit. When a rotating dc machine is operated as a motor, the generated CEMF (V_c) opposes the armature current and is *less than* the voltage applied to the machine (V_t). This is consistent with Lenz's law. These relationships may be summarized as follows:

$$\text{For dc motors:} \quad V_a = V_c + I_a R_a$$
$$\text{For dc generators:} \quad V_g = V_a + I_a R_a$$

where V_a = applied voltage (also called "terminal voltage")
           V_c = CEMF developed in a motor armature circuit
           V_g = voltage generated or induced into a generator armature
    $I_a R_a$ = voltage drop across the armature circuit

The CEMF (V_c) is expressed as $V_c = V_a - I_a R_a$.

The torque developed by a motor is dependent on armature current. The value of armature current may be calculated by the formula

$$I_a = \frac{V_a - V_c}{R_a}$$

The force (F) per conductor developed due to motor action is expressed as

$$F = \frac{BIl}{1.13} \times 10^{-7} \text{ lb}$$

where B = flux density in lines of force per square inch
        I = current in amperes
        l = length of the conductor
   1.13 = a constant

For example, if a conductor is 10 in. long, current flow through it is 15 A, and the magnetic field strength is 30,000 lines/in², the force (F) developed by the conductor is

$$F = \frac{BIl}{1.13} \times 10^{-7}$$

$$= \frac{30,000 \text{ lines/in}^2 \times 15\text{A} \times 10 \text{ in.}}{1.13} \times 10^{-7}$$

$$= \frac{4,500,000}{1.13} \times 10^{-7} = 3.98 \text{ lb}$$

Torque developed in motors is a force that produces mechanical energy in the form of rotation. It is the product of force and distance and is commonly expressed in units of foot-pounds (ft-lb), ounce-inches (oz-in.), or kilogram-meters (kg-m).

The calculation of torque of a dc motor may be accomplished by considering the percentage of armature conductors that lie directly adjacent to a field pole. This calculation is based on the assumption that the armature conductors which are positioned directly adjacent to a field pole contribute in an equal manner to the production of torque. The average force (F_{avg}) is found by the equation $F_{avg} = F_c \times C_a$, where F_c is the average force per armature conductor adjacent to a field pole in pounds and C_a is the number of armature conductors. The average torque (T_{avg}) of the machine (in ft-lb) may be computed as $T_{avg} = F_{avg} \times r = F_c \times C_a \times r$, where r is the radial distance to the axis of rotation in feet.

For example, the armature of a dc motor has 200 conductors, a diameter of 6 in., and conductor length of 12 in. The flux density of the field is 30,000 lines/in^2 and 75% of the armature conductors are adjacent to a field winding. The armature current flow is 10 A. The force per conductor (F_c) is found as follows:

$$F_c = \frac{Bll}{1.13} \times 10^{-7}$$

$$= \frac{30,000 \text{ lines/in}^2 \times 10 \times 12}{1.13} \times 10^{-7}$$

$$= \frac{3,600,000}{1.13} \times 10^{-7} = 3.18 \text{ lb}$$

The average force (F_{avg}) tending to rotate the armature may then be calculated as

$$F_{avg} = F_c \times C_a$$

$$= 3.18 \times (200 \times 0.75)$$

$$= 477 \text{ lb}$$

The average torque (T_{avg}) is then found as

$$T_{avg} = F_{avg} \times r$$

$$= 477 \text{ lb} \times 0.25 \text{ ft} \quad (3 \text{ in.} = \tfrac{1}{2} \text{ of } 6 \text{ in.} = 0.25 \text{ ft})$$

$$= 119.25 \text{ ft-lb}$$

The torque of a dc motor field may also be determined by the formula

$$T = K\phi I_A \qquad \text{ft-lb}$$

where K = a constant for a specific machine
  ϕ = magnetic flux per pole in lines/in^2
  I_A = armature current in amperes

The equation above shows that as either magnetic flux or armature current values increase, torque increases in direct proportion.

Horsepower rating of motors. The horsepower rating of a motor is based on the amount of torque produced at the rated full-load values. Horsepower, which is the usual method of rating motors, can be expressed mathematically as

$$\text{hp} = \frac{2\pi NT}{33,000}$$

$$= \frac{NT}{5252}$$

where hp = horsepower rating
  2π = a constant
  N = speed of the motor in r/min
  T = torque developed by the motor in foot-pounds

DC MOTOR CHARACTERISTICS

Motors that operate from dc power sources have many applications in industry when speed control is desirable. Dc motors are almost identical in construction to dc generators. They are also classified in a similar manner as series, shunt, or compound machines, depending on the method of connecting the armature and field windings. Also, permanent-magnet dc motors are used for certain applications. A typical dc motor is shown in Figure 8-3. The operational characteristics of all dc motors can be generalized by referring to Figure 8-4. Most electric motors exhibit characteristics similar to those shown in the block diagram. When discussing dc motor characteristics, it is necessary to be familiar with the following terms: *load, speed, counterelectromotive force* (CEMF), *armature current,* and *torque*. The amount of mechanical load applied to the shaft of a motor determines its operational characteristics. As the mechanical load is increased, the speed of a motor tends to decrease. As the speed decreases, the voltage induced into the conductors of the motor due to generator action (CEMF) decreases. The generated voltage, or counterelectromotive force, depends on the number of rotating conductors and the speed of rotation. Therefore, as the speed of rotation decreases, so does the CEMF.

The counterelectromotive force generated by a motor is in opposition to the supply voltage. Since the CEMF is in opposition to the supply voltage, the actual working voltage of a motor will increase as the CEMF decreases. When the working voltage increases, more current will flow through the armature windings. Since the torque of a motor is directly proportional to the armature current, the torque will increase as the armature current increases.

To discuss briefly the opposite situation, if the mechanical load connected to the motor decreases, speed tends to increase. An increase in speed causes an increase in the CEMF. Since the CEMF is in opposition to the supply voltage, as the CEMF increases, the armature current decreases. A decrease in armature current causes a decrease in torque. Torque varies directly with changes in load. As the load on a motor

Figure 8-3 Typical direct-current motor. (Courtesy of RAE Corp.)

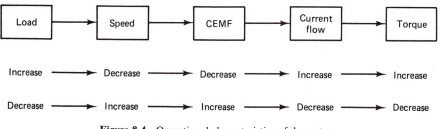

Figure 8-4 Operational characteristics of dc motors.

is increased, its torque also increases to try to meet the increased load requirement. However, the current drawn by a motor also increases when the load is increased.

The presence of a CEMF to oppose the armature current is very important in motor operation. The lack of any CEMF when a motor is being started explains why motors draw a very large initial starting current compared to their running current when full speed is reached. Maximum armature current flows when there is no CEMF. As the CEMF increases, the armature current decreases. Thus resistances in series with the armature circuit are often used to compensate for the lack of CEMF and to reduce the starting current of a motor. After the motor has reached full speed, these resistances are bypassed by automatic or manual switching systems in order to allow the motor to produce maximum torque. Keep in mind that the armature current, which directly affects torque, can be expressed as

$$I_A = \frac{V_T - V_C}{R_A}$$

where I_A = armature current in amperes
V_T = terminal voltage of the motor in volts
V_C = CEMF generated by the motor in volts
R_A = armature resistance in ohms

Another dc motor characteristic that should be discussed is *armature reaction.* Armature reaction was discussed in relation to dc generators in Chapter 5. This effect distorts the main magnetic field in dc motors as well as in dc generators. Similarly, the brushes of a dc motor can be shifted to counteract the effect of armature reaction. However, interpoles or compensating windings are ordinarily used to control armature reaction in dc motors.

Speed characteristics of dc motors.   The most desirable characteristic of dc motors is their speed-control capability. By varying the applied dc voltage with a rheostat, speed can be varied from zero to the maximum r/min of the motor. Some types of dc motors have more desirable speed characteristics than do others. For this reason, comparative speed regulation for different types of motors can be calculated. Speed regulation is expressed as

$$\% \ R = \frac{S_{NL} - S_{FL}}{S_{FL}} \times 100$$

where $\% \ R$ = percentage of speed regulation
S_{NL} = no-load speed in r/min
S_{FL} = rated full-load speed in r/min

Good speed regulation (low $\% \ R$) results when a motor has nearly constant speeds under varying load situations.

TYPES OF DC MOTORS

The types of commercially available dc motors basically fall into four categories: (1) permanent-magnet dc motors, (2) series-wound dc motors, (3) shunt-wound dc motors, and (4) compound-wound dc motors. Each of these motors has different characteristics due to its basic circuit arrangement and physical properties.

Permanent-magnet dc motors.   The permanent-magnet dc motor, shown in Figure 8-5, is constructed in the same manner as its dc generator counterpart that was discussed in Chapter 5. The permanent-magnet motor is used for low-torque applications. When this type of motor is used, the dc power supply is connected directly to the armature conductors through the brush/commutator assembly. The magnetic field is produced by permanent magnets mounted on the stator. The rotor of permanent magnet motors is a wound armature.

This type of motor ordinarily uses either alnico or ceramic permanent magnets rather than field coils. The alnico magnets are used with high-horsepower applications. Ceramic magnets are ordinarily used for low-horsepower slow-speed motors. Ceramic magnets are highly resistant to demagnetization, yet they are relatively low in magnetic-flux level. The magnets are usually mounted in the motor frame and then magnetized prior to the insertion of the armature. Two permanent-magnet dc motors are shown in Figure 8-6.

The permanent-magnet motor has several advantages over conventional types of

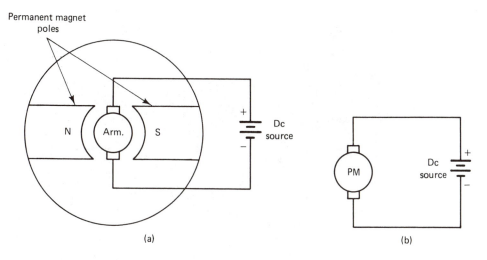

Figure 8-5 Permanent-magnet dc motor: (a) pictorial diagram; (b) schematic diagram.

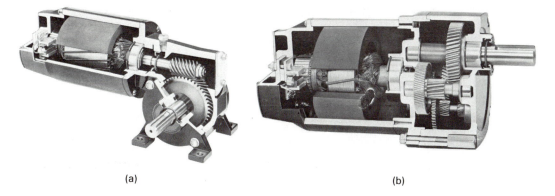

Figure 8-6 Permanent-magnet dc motors: (a) right-angle gear motor; (b) parallel-shaft gear motor. (Courtesy of Bodine Electric Co.)

dc motors. One advantage is reduced operational cost. The speed characteristics of the permanent-magnet motor are similar to those of the shunt-wound dc motor. The direction of rotation of a permanent-magnet motor can be reversed by reversing the two power lines.

Series-wound dc motors. The manner in which the armature and field circuits of a dc motor are connected determines its basic characteristics. Each of the types of dc motors are similar in construction to the type of dc generator that corresponds to it. The only difference, in most cases, is that the generator acts as a voltage source while the motor functions as a mechanical power conversion device. A typical dc motor with a wound armature and field windings is shown in Figure 8-7.

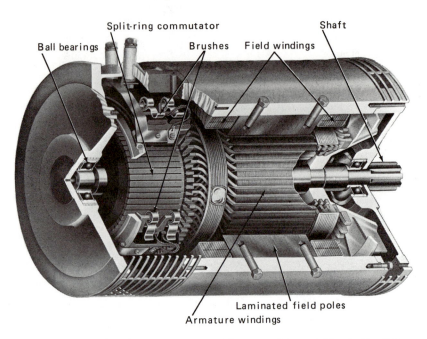

Figure 8-7   Typical direct-current motor. (Courtesy of General Electric Co., DC Motor and Generator Dept.)

Figure 8-8   Series-wound dc motor: (a) pictorial diagram; (b) schematic diagram.

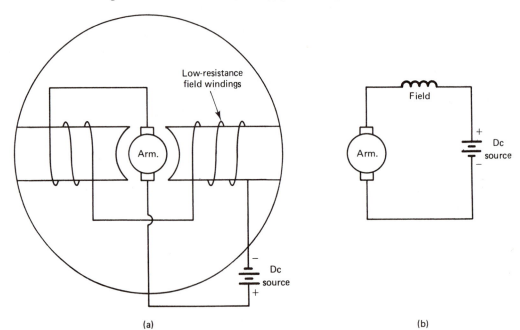

(a)

(b)

The series-wound motor, shown in Figure 8-8, has the armature and field circuits connected in a series arrangement. There is only one path for current to flow from the dc voltage source. Therefore, the field is wound of relatively few turns of large-diameter wire, giving the field a low resistance. Changes in load applied to the motor shaft causes changes in the current through the field. If the mechanical load increases, the current also increases. The increased current creates a stronger magnetic field. The speed of a series motor varies from very fast at no load to very slow at heavy loads. Since large currents may flow through the low-resistance field, the series motor produces a high-torque output. Series motors are used where heavy loads must be moved and speed regulation is not important. A typical application is for automobile starter motors.

Shunt-wound dc motors.   Shunt-wound dc motors are more commonly used than any other type of dc motor. As shown in Figure 8-9, the shunt-wound dc motor has field coils connected in parallel with its armature. This type of dc motor has field coils which are wound of many turns of small-diameter wire and have a relatively high resistance. Since the field is a high-resistant parallel path of the circuit of the shunt motor, a small amount of current flows through the field. A strong electromagnetic field is produced due to the many turns of wire that form the field windings.

A large majority (about 95%) of the current drawn by the shunt motor flows in the armature circuit. Since the field current has little effect on the strength of the field,

Figure 8-9   Shunt-wound dc motor: (a) pictorial diagram; (b) schematic diagram.

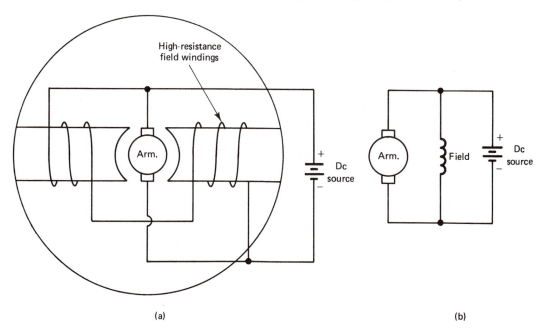

(a)   (b)

motor speed is not affected appreciably by variations in load current. The relationship of the currents that flow through a dc shunt motor is as follows:

$$I_T = I_A + I_F$$

where I_T = total current drawn from the power source
 I_A = armature current
 I_F = field current

The field current may be varied by placing a variable resistance in series with the field windings. Since the current in the field circuit is low, a low-wattage rheostat may be used to vary the speed of the motor due to the variation in field resistance. As field resistance increases, field current will decrease. A decrease in field current reduces the strength of the electromagnetic field. When the field flux is decreased, the armature will rotate *faster,* due to reduced magnetic-field interaction. Thus the speed of a dc shunt motor may be easily varied by using a field rheostat.

The shuntwound dc motor has very good speed regulation. The speed does decrease slightly when the load increases due to the increase in voltage drop across the armature. Due to its good speed regulation characteristic and its ease of speed control, the dc shunt motor is commonly used for industrial applications. Many types of variable-speed machine tools are driven by dc shunt motors.

Compound-wound dc motors.   The compound-wound dc motor shown in Figure 8-10, has two sets of field windings, one in series with the armature and one in parallel. This motor combines the desirable characteristics of the series- and shunt-wound motors. There are two methods of connecting compound motors: cumulative and differential. A *cumulative* compound motor has series and shunt fields that *aid* each other. *Differential* compound dc motors have series and shunt fields that

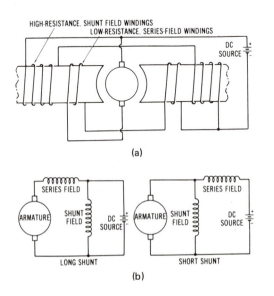

Figure 8-10 Compound-wound dc motors: (a) pictorial diagram; (b) schematic diagrams.

oppose each other. There are also two ways in which the series windings are placed in the circuit. One method is called a *short-shunt* (see Figure 8-10), in which the shunt field is placed across the armature. The *long-shunt* method has the shunt field winding placed across both the armature and the series field (see Figure 8-10).

Compound motors have high torque similar to a series-wound motor, together with good speed regulation similar to a shunt motor. Therefore, when good torque and good speed regulation are needed, the compound-wound dc motor can be used. A major disadvantage of a compound-wound motor is its expense.

DC MOTOR ROTATION REVERSAL

A dc motor is designed so that its shaft will rotate in either direction. It is a very simple process to reverse the direction or rotation of any dc motor. By reversing the relationship between the connections of the armature windings and field windings, reversal of rotation is achieved. Usually, this is done by changing the terminal connections where the power source is connected to the motor. Four terminals are ordinarily used for interconnection purposes. They may be labeled A1 and A2 for the armature connections and F1 and F2 for the field connections. If either the armature connections or the field connections are reversed, the rotation of the motor will reverse. However, if both are reversed, the motor shaft will rotate in its original direction, since the relationship between the armature and field windings would still be the same.

COMPARISON OF DC MOTOR CHARACTERISTICS

The characteristics of dc motors should be considered when selecting motors for particular applications. Figures 8-11 and 8-12 show comparative graphs that illustrate the relative torque and speed characteristics of dc motors.

Torque relationships.   A comparative set of torque versus armature current curves for dc motors is shown in Figure 8-11. The effect of increased mechanical load on the shaft of each type of motor can be predicted.

Series-wound dc motors have equal armature current, field current, and load current ($I_A = I_F = I_L$). The magnetic flux (ϕ) produced by the field windings is proportional to the armature current (I_A). The torque produced by a series-wound motor with low values of I_A is less than other motors due to lack of field flux development. However, at rated full-load armature current the torque is greater than other types of dc motors.

Shunt-wound dc motors have a fairly constant magnetic field flux due to the high-resistant field circuit. An almost linear torque versus armature current curve is a characteristic of shunt-wound dc motors. Since torque is directly dependent on armature current, as I_A increases, torque increases in direct proportion.

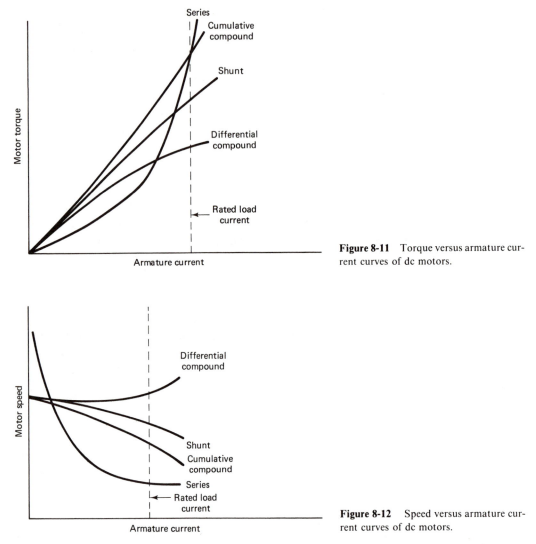

Figure 8-11 Torque versus armature current curves of dc motors.

Figure 8-12 Speed versus armature current curves of dc motors.

Compound-wound dc motors are of two general types: cumulative and differential. *Cumulative* compound dc motors have series and shunt field windings which *aid* each other in the production of an overall magnetic field. In this type of motor circuit the general torque equation is: $T = K(\phi_S + \phi_P)I_A$, where ϕ_S is the series field flux and ϕ_P is the parallel (shunt) field flux. The series field flux increases as I_A increases and the shunt field flux remains fairly constant. Therefore, the torque curve (see Figure 8-11) for a cumulative compound dc motor is always higher than that of a similar shunt-wound dc motor.

The *differential* compound dc motor has series and shunt fields which *oppose* each other in the production of an overall magnetic field. The general torque equation for this type of motor is $T = K(\phi_S - \phi_P)I_A$. The value of ϕ_S depends primarily on armature current and ϕ_P is fairly constant. Since these magnetic flux values oppose each other, the torque curve of a differential compound motor is always less than that of a shunt-wound dc motor.

Speed relationships. A comparative set of speed versus armature current curves for dc motors is shown in Figure 8-12. The effect of increased armature current on motor speed for each type of motor can be predicted.

Series-wound dc motors have speed characteristics which conform to the basic speed equation:

$$S = \frac{V_A - I_A \times (R_A + R_F)}{K\phi}$$

where V_A = rated armature voltage
$\quad I_A$ = armature current
$\quad R_A$ = armature resistance
$\quad R_F$ = resistance of the series field winding
$\quad K$ = a constant
$\quad \phi$ = magnetic flux of the field windings

Since the magnetic flux produced by the series field is proportional to armature current flow, the speed equation may be written as

$$S = K \frac{V_A - I_A \times (R_A + R_F)}{I_A}$$

With a small mechanical load applied to the shaft of a series-wound dc motor, the armature current (I_A) is very low. A low value of I_A causes the numerator of the speed equation to be large and the denominator small. The result is an extremely high speed. At no-load, the I_A is small, field flux (ϕ) is small, and speed is excessively high. Series motors have a "runaway" speed characteristic at no-load. Thus a load should always be connected to a series motor shaft to prevent the motor speed from becoming so high as to damage the motor. When the load on a series-wound dc motor increases, the numerator of the speed equation decreases faster than the denominator increases and the speed drops appreciably.

Shunt-wound dc motors have speed characteristics based on the speed equation:

$$S = \frac{V_A - (I_A R_A)}{K\phi}$$

The magnetic field flux (ϕ) of a shunt-wound dc motor is practically constant. As mechanical load applied to the shaft of shunt-wound dc motor increases, the CEMF decreases, and speed decreases slightly due to the difference between V_A and the product of $I_A R_A$. The speed of a shunt-wound dc motor is considered to be almost constant.

Compound-wound dc motors of the cumulative type have the following speed equation:

$$S = K \ \frac{V_A - I_A \times (R_A + R_F)}{\phi_S + \phi_P}$$

As the load applied to the motor increases, I_A increases, the flux produced by the series field (ϕ_S) increases, the CEMF decreases, and the numerator decreases at an even faster rate. The speed thus decreases with increases in load.

The differential compound motor has a speed equation of

$$S = K \ \frac{V_A - I_A \times (R_A + R_F)}{\phi_P - \phi_S}$$

The parallel and series field windings develop an overall magnetic flux in which the two (ϕ_P and ϕ_S) are in opposition. As load increases, the numerator of the speed equation decreases, but the denominator decreases more rapidly. Thus speed *increases* with increases in load. For this reason, differential compound motors are seldom used.

DYNAMOTORS

Another type of dc motor is called a *dynamotor*. This motor, depicted in Figure 8-13, converts dc voltage of one value to dc voltage of another value. It is actually a motor-generator housed in one unit. The armature has two separate windings. One winding is connected to the commutator of the motor section and the other winding is connected to the commutator of a generator unit. A magnetic field, developed by either permanent magnets or electromagnetic windings, surrounds the armature assembly. Since the magnetic field remains relatively constant, the generator voltage output depends on the ratio of the number of motor windings to the number of generator windings. For instance, if there are twice as many generator windings as motor windings, the generated dc voltage output will be twice the value of the dc voltage that is input to the motor section of the dynamotor.

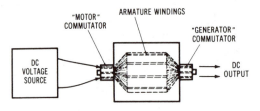

Figure 8-13   Dynamotor construction.

BRUSHLESS DC MOTORS

The use of transistors has resulted in the development of brushless dc motors that have neither brushes nor commutator assemblies. Instead, they make use of solid-state switching circuits. The major problem with most dc motors is the low reliability of the

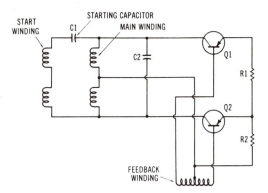

START
WINDING

STARTING CAPACITOR

C1

MAIN WINDING

C2

Q1

R1

Q2

R2

FEEDBACK
WINDING

Figure 8-14 Brushless dc motor circuit.

commutator/brush assembly. The brushes have a limited life and cause the commutator to wear. This wearing produces brush dust which can cause other maintenance problems.

Although some brushless dc motors use other methods, the transistor-switched motor is the most common (see Figure 8-14). The motor itself is actually a single-phase ac permanent-capacitor induction motor with a center-tapped main winding. Transistors, which are operated by an oscillator circuit, conduct alternately through the paths of the main winding. The oscillator circuit requires a feedback winding wound into the stator slots, which generates a control voltage to determine the frequency. A capacitor (C2) is placed across the main winding to reduce voltage peaks and to keep the frequency of the circuit at a constant value.

The main disadvantage of this motor is its inability to develop a very high starting torque. As a result, it is suitable only for driving very low torque loads. When used in a low-voltage system, this motor is not very efficient. Also, since only half of the main winding is in use at any instant, copper losses are relatively high. However, the advantages outweigh this disadvantage for certain applications. Since there are no brushes and commutator, motor life is limited mainly by the bearing. With proper lubrication, a brushless dc motor can be used for an indefinite period. Also, the motor frequency and thus the speed can be adjusted by varying the oscillator circuit.

DC MOTOR STARTING CIRCUITS

When dc motors are initially started, a problem exists due to lack of CEMF in the armature circuit. Excessive starting current flows through the low resistance of the armature conductors. Once CEMF builds up as speed increases, the problem of high starting current is less evident.

The example below illustrates the need for dc motor starting circuits which limit current flow until speed of rotation of the motor has increased. Assume that a 240-V dc shunt motor has an armature resistance of $0.2\,\Omega$ and a rated full-load armature current of 30 A. The starting current (I_S) equals

$$I_S = \frac{V_A}{R_A} = \frac{240\ \text{V}}{0.2\ \Omega} = 120\ \text{A} \qquad \text{(with CEMF} = 0\ \text{V)}$$

179

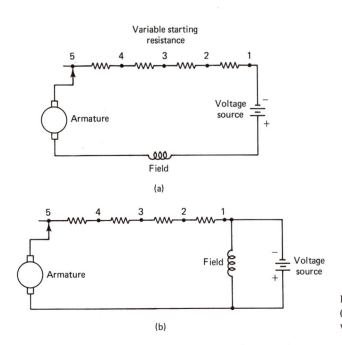

Figure 8-15 Dc motor starting circuits: (a) series-wound motor control; (b) shunt-wound motor control.

The percent of rated full-load armature current is

$$\% \text{ full-load} = \frac{120 \text{ A}}{30 \text{ A}} \times 100 = 400\%$$

Dc motor starting circuits employ resistances connected in series with the power lines to limit starting current. Since motor torque is dependent on current flow, the resistances are used only during the starting cycle of the motor. They are either manually or automatically removed from the motor circuit once operating speed is reached. Starting resistances are often used in steps as shown in Figure 8-15. A decreasing value of starting resistance is needed as motor speed increases. The value of starting resistance required for various values of rated full-load current may be calculated as follows. Assume that the 240-V dc motor of the previous example is used. The starting resistance (R_{ST}) required in series with the armature to limit starting current to 150% of rated full-load armature current is

$$R_{\text{total}} = \frac{V_A}{I_A} = \frac{240 \text{ V}}{150\% \text{ of } 30 \text{ A}} = \frac{240 \text{ V}}{45 \text{ A}} = 5.33 \text{ }\Omega$$

$$R_{ST} = R_{\text{total}} - R_A$$

$$= 5.33 \text{ }\Omega - 0.2 \text{ }\Omega$$

$$= 5.13 \text{ }\Omega$$

Ordinarily, dc motors use starting circuits that limit armature current during initial starting to a value greater than rated full-load current. This allows a motor to produce greater starting torque.

REVIEW

8.1. List some applications of electric motors.

8.2. Discuss "motor action" or torque development in a motor.

8.3. What is the "right-hand motor rule"?

8.4. What is the relationship of applied voltage and generated voltage in dc motors and generators?

8.5. How is force developed by motor action calculated?

8.6. How is the average torque of a motor calculated?

8.7. How is the horsepower of a motor calculated?

8.8. What is the relationship of load, speed, CEMF, armature current, and torque of a dc motor?

8.9. How does armature reaction affect dc motor operation?

8.10. What is speed regulation of a motor?

8.11. What are the types of dc motors?

8.12. Discuss the construction of each of the major types of dc motors.

8.13. What are the major operational characteristics of each major type of dc motor?

8.14. How is speed calculated for the following types of dc motors: **(a)** series-wound, **(b)** shunt-wound, and **(c)** cumulative compound?

8.15. Discuss the operation of a dynamotor.

8.16. Discuss the operation of a brushless dc motor.

8.17. What is the main problem of starting dc motors?

8.18. What method is ordinarily used to start dc motors?

PROBLEMS

8.1. A dc motor has 100 V applied, armature resistance of 1.8 Ω, and an armature current of 32 A. What is the value of the CEMF developed?

8.2. A dc motor has 90 V applied, a CEMF of 33 V, and an armature resistance of 3.2 Ω. What is the value of the armature current?

8.3. An armature conductor of a dc motor is 8 in. long, draws 7.5 A of current, and has a surrounding magnetic field strength of 24,000 lines/in^2. What is the force developed by the conductor?

8.4. If the armature of Problem 8.3 has 120 conductors, a diameter of 6 in., and 80% of the armature conductors adjacent to a field winding, what is:
(a) Average force?
(b) Average torque?

8.5. What is the horsepower of a motor running at 1750 r/min and having a torque of 25 ft-lb?

8.6. If the no-load speed of a motor is 3550 r/min and the full-load speed is 3450 r/min, what is its speed regulation?

8.7. A shunt-wound dc motor converts 5000 W when connected to a 120-V line. If its field resistance equals 100 Ω, find:

(a) Field current

(b) Line current

(c) Armature current

8.8. If a shunt-wound dc motor draws 60 A from a 240-V source, its field resistance is 40 Ω, and its armature resistance is 0.5 Ω, find:

(a) Field current   (b) Armature current

(c) CEMF   (d) Total power

8.9. A 240-V dc shunt motor has an armature resistance of 0.25 Ω and armature current of 30 A. Calculate:

(a) Generated voltage

(b) Armature power

8.10. A 120-V dc shunt motor has an armature resistance of 0.5 Ω and a rated full-load current of 20 A. Calculate:

(a) Starting current

(b) Percent full-load current

(c) Limiting resistance required for 200% of full-load current

(d) Resistance required for 150% of full-load current

NINE

Single-Phase
Alternating-Current Motors

Single-phase ac motors are common for industrial as well as commercial and residential usage. They operate from a single-phase ac power source. There are three basic types of single-phase ac motors. These types are *universal* motors, *induction* motors, and *synchronous* motors.

UNIVERSAL MOTORS

Universal motors can be powered by either ac or dc power sources. The universal motor, illustrated in Figure 9-1, is constructed in the same way as a series-wound dc motor; however, it is designed to operate with either ac or dc applied. The series-wound motor is the only type of dc motor that will operate with ac applied. Shunt-wound motors have windings with inductance values that are too high to function with ac applied. However, series motors have windings that have low inductance (a few turns of large-diameter wire), and therefore offer a low impedance to the flow of alternating current. The universal motor is one type of ac motor that has concentrated (or salient) field windings. These field windings are similar to those of all dc motors.

The operational principle of the universal motor with ac applied involves the instantaneous change of both field and armature polarities. Since the field windings have low inductance, the reversals of field polarity brought about by the changing nature of alternating current also create reversals of current direction through the armature conductors at the proper time intervals. The universal motor operates in the same manner as a series-wound dc motor except that the field polarity and the direction of armature current change at a rate of 120 times per second when connected

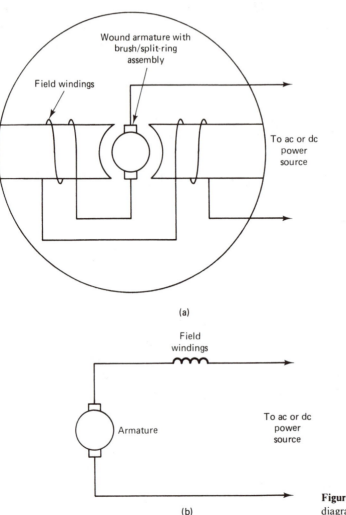

(a)

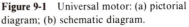

(b)

Figure 9-1 Universal motor: (a) pictorial diagram; (b) schematic diagram.

to a 60-Hz ac source. The speed and torque characteristics of universal motors are similar to those of dc series-wound motors. Applications of universal motors in industry are mainly for portable tools and small motor-driven equipment. Home applications include mixers, blenders, and electric drills.

INDUCTION MOTORS

Single-phase ac induction motors are illustrated in Figure 9-2. The coil symbols along the stator represent the field coils of an induction motor. These coils are energized by an ac source; therefore, their instantaneous polarity changes 120 times per second when 60-Hz ac is applied.

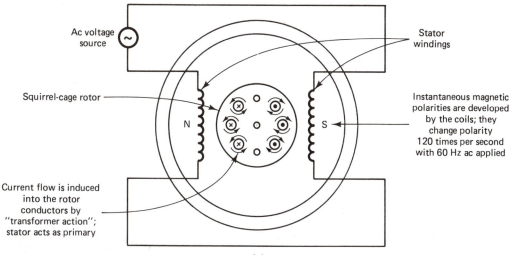

Figure 9-2   Single-phase ac induction motor: (a) schematic diagram; (b) assembly, [(b), Courtesy of Leeson Electric Corp.)]

Induction motors have a solid rotor, referred to as a *squirrel-cage* rotor. This type of rotor, which is illustrated in Figure 9-3, has large-diameter copper conductors soldered at each end to a circular connecting plate. This plate actually short-circuits the individual conductors of the rotor. When current flows in the stator windings, a current is induced in the rotor. This current is developed due to "transformer action" between the stator and rotor. The stator, which has ac voltage applied, acts as a

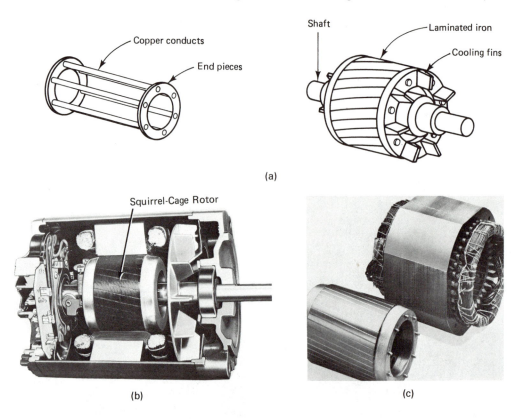

Figure 9-3 (a) Pictorial representation of a squirrel-cage rotor; (b) cutaway of an induction motor showing a squirrel-cage rotor; (c) squirrel-cage rotor and stator of a single-phase ac induction motor. [(b), Courtesy of Franklin Electric Co.; (c), courtesy of Leeson Electric Corp.]

transformer primary. The rotor is similar to a transformer secondary since current is induced into it.

Since the stator polarity changes in step with the applied ac, it develops a rotating magnetic field. The rotor becomes instantaneously polarized due to the induced current flow through the short-circuited copper conductors. The rotor will, therefore, tend to rotate in step with the revolving magnetic field of the stator. If some method of initially starting the rotation is used, the rotor will continue to rotate. However, due to inertia, a rotor must be initially put into motion by some auxiliary method.

It should be pointed out that the speed of an ac induction motor is based on the speed of the rotating magnetic field and the number of stator poles that the motor has. The speed of the rotor will never be as high as the speed of the rotating stator field. If the two speeds were equal, there would be no relative motion between the rotor and stator and, therefore, no induced rotor current and no torque would be developed. The rotor speed of an induction motor is always somewhat less than the rotating stator field developed by the applied ac voltage.

The speed of the rotating stator field can be expressed as

$$S = \frac{f \times 120}{N}$$

where S = speed of the rotating stator field in r/min
       f = frequency of the applied ac voltage in hertz
     N = number of poles of the stator windings
   120 = a conversion constant

A two-pole motor operating from a 60-Hz source would have a stator speed of 3600 r/min. The stator speed is also referred to as the *synchronous speed* of a motor.

Assume that an induction motor has four stator poles and operates on 60 Hz. Its synchronous (stator) speed is

$$S = \frac{f \times 120}{N} = \frac{60 \times 120}{4} = \frac{7200}{4} = 1800 \text{ r/min}$$

The difference between the revolving stator speed of an induction motor and the rotor speed is called *slip*. The rotor speed must lag behind the revolving stator speed in order to develop torque. The more the rotor speed lags behind, the more torque is developed. Slip is expressed mathematically as

$$\% \text{ slip} = \frac{S_S - S_R}{S_S} \times 100$$

where S_S = synchronous (stator) speed in r/min
     S_R = rotor speed in r/min

As the rotor speed approaches the stator speed, the percentage of slip becomes smaller.

Assume that an induction motor has a synchronous speed of 3600 r/min and a rotor operating speed of 3400 r/min. The percentage of slip is

$$\% \text{ slip} = \frac{S_S - S_R}{S_S} \times 100$$

$$= \frac{3600 - 3400}{3600} = \frac{200}{3600} = 0.055 \times 100 = 5.5\%$$

Another factor, referred to as *rotor frequency,* affects the operational characteristics of an induction motor under load. As the load on the shaft of the motor increases, the rotor speed tends to decrease. The stator speed, however, is unaffected. When a two-pole induction motor connected to a 60-Hz source operates at 10% slip, the rotor frequency would equal 360 r/min (3600 r/min × 10%). Functionally, this means that a revolving stator field passes across a rotor conductor 360 times per minute. Current is induced into a rotor conductor each time the stator field revolves past the conductor. As slip is increased, more current is induced into the rotor, causing more torque to be developed. The rotor frequency depends on the amount of slip and can be expressed as

$$f_r = f_s \times \text{slip}$$

where f_r = frequency of the rotor current in r/min
       f_s = frequency of the stator current in r/min

and slip is expressed as a decimal. Rotor frequency affects the operational characteristics of induction motors.

     Single-phase ac induction motors are classified according to the method used for starting. Common types of single-phase ac induction motors include: split-phase motors, capacitor motors, shaded-pole motors, and repulsion motors.

     Split-phase motors.   The split-phase ac induction motor, illustrated in Figure 9-4, has two sets of stator windings. One set, called the *run windings,* is connected directly across the ac line. The other set, called the *start windings,* is also connected across the ac line. However, the start windings are connected in series with a centrifugal switch mounted on the shaft of the motor. The centrifugal switch is in the closed position when the motor is not rotating.

     Before discussing the functional principle of the split-phase ac motor, an understanding of how rotation is developed by an ac motor is important. Refer to Figure 9-5. In Figure 9-5(a), a two-pole stator with single-phase ac applied is shown. For purposes of this discussion, a permanent magnet is placed inside the stator to represent the squirrel-cage rotor of an induction motor. At time t_0 of the ac sine wave, no stator field is developed. Time interval t_1 will cause a stator field to be produced. Assume a north polarity on the right pole of the stator and a south polarity on the left pole. These polarities will cause the rotor to align itself horizontally based on the laws of magnetic attraction. At time t_2, the stator poles become demagnetized and then begin to magnetize in the opposite direction. At time interval t_3, the stator poles are magnetized in the opposite direction. The rotor will now align itself horizontally in the opposite direction. This effect will continue at a rate of 120 polarity changes per second if 60-Hz ac is applied to the stator. The rotor would not start unless it was positioned initially to be drawn toward a pole piece. Therefore, some starting method must be used for single-phase ac motors, since they are not self-starting.

     Assume that we have a two-phase situation as shown in Figure 9-5(b). We now have two sets of stator windings with one phase connected to each set. Two-phase voltage is, of course, not produced by power companies in the United States; however, this example will show the operational principle of a split-phase motor. In the two-phase voltage diagram, when one phase is at minimum value, the other is at maximum value.

     At time interval t_1, phase 1 is maximum while phase 2 is minimum. Assume that the right stator pole becomes a north polarity and the left stator pole becomes a south

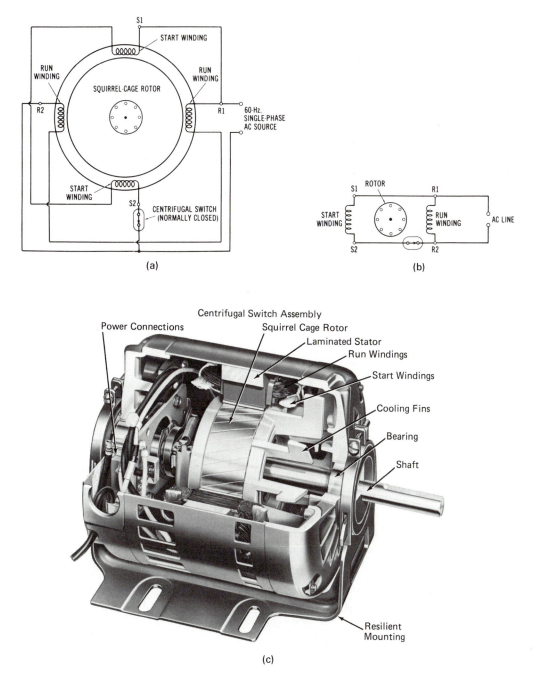

(a)

(b)

(c)

Figure 9-4 Split-phase ac induction motor: (a) pictorial diagram; (b) schematic diagram; (c) cutaway view. [(c), Courtesy of Marathon Electric.]

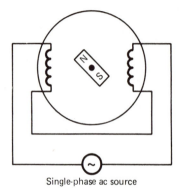
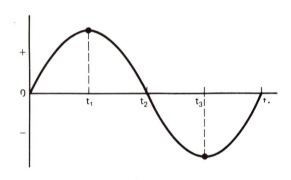

Single-phase ac source

(a)

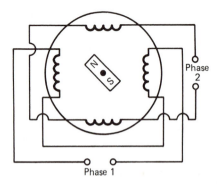
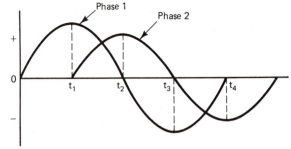

(b)

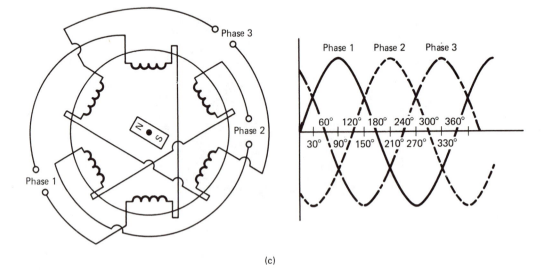

(c)

Figure 9-5 Method of rotation development in ac motors: (a) single-phase situation; (b) two-phase situation; (c) three-phase situation.

polarity. The rotor will align itself horizontally since no polarity is developed in the vertical poles at this time. Now, as we progress to time t_2, phase 1 is minimum and phase 2 is maximum. Assume that the upper stator pole becomes a south polarity and the bottom stator pole becomes a north polarity. The rotor will now align itself vertically, by moving 90°. At time t_3, phase 1 becomes maximum in the opposite direction and phase 2 is minimum. This time interval results in a north pole on the left and a south pole on the right. Thus the rotor moves 90° farther in a clockwise direction. This effect will continue as two-phase voltage is applied to the stator poles. We can see from the two-phase situation that a direction of rotation is established by the relationship of phase 1 and phase 2, and that an ac motor with two-phase voltage applied would be self-starting. The same is true for a three-phase situation, which is illustrated in Figure 9-5(c). Rotation of the rotor would result due to the 120° phase separation of the three-phase voltage applied to the stator poles. Three-phase induction motors are therefore self-starting with no auxiliary starting method required.

Referring back to the split-phase motor of Figure 9-4, it can be seen that the purpose of the two sets of windings is to establish a simulated two-phase condition to start the motor. The single-phase voltage applied to this motor is said to be "split" into a two-phase current. A rotating magnetic field is created by phase splitting. The start winding of the split-phase motor is made of relatively few turns of small-diameter wire, giving it a high resistance and a low inductance. The run winding is wound with many turns of large-diameter wire, causing it to have a lower resistance and a higher inductance. We know that inductance in an ac circuit causes the current to lag the applied ac voltage. The more inductance present, the greater is the lag in current.

When single-phase ac is applied to the stator of a split-phase induction motor, the situation illustrated in Figure 9-6 will result. Notice that the current in the start winding lags the applied voltage due to its inductance. However, the current in the run winding lags by a greater amount due to its higher inductance. The phase separation of the

Figure 9-6   Relationship of voltage and current in a split-phase ac induction motor.

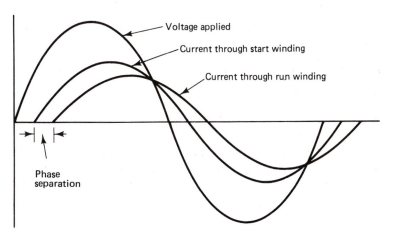

Voltage applied

Current through start winding

Current through run winding

Phase
separation

currents in the start and run windings creates a two-phase situation. The phase displacement, however, is usually around 30° or less, which gives the motor a low starting torque since this phase separation does not nearly approach the 90° separation of two-phase voltage.

When the split-phase ac induction motor reaches about 80% of its normal operating speed, the centrifugal switch will open the start winding circuit, since it is no longer needed. The removal of the start winding minimizes energy losses in the machine and prevents the winding from overheating. When the motor is turned off and its speed reduces, the centrifugal switch closes to connect the start winding back into the circuit.

Split-phase motors are fairly inexpensive compared to other types of single-phase motors. They are used where low torque is required to drive mechanical loads such as small machinery. Figure 9-7 shows the classes of induction motors according to starting characteristics.

Class	Starting characteristics
A	Moderate torque, moderate starting current
B	Moderate torque, low starting current
C	High torque, low starting current
D	High slip
E	Low starting torque, moderate starting current
F	Low starting torque, low starting current

Figure 9-7 Classes of ac induction motors.

Capacitor motors.   Capacitor motors are an improvement over the split-phase ac motor. Capacitor-start, single-phase induction motors are shown in Figure 9-8. Notice that this diagram is the same as the split-phase motor except that a capacitor is placed in series with the start winding. The purpose of the capacitor is to cause the current in the start winding to lead (rather than lag) the applied voltage.

This situation is illustrated in Figure 9-9. The current in the start winding now leads the applied voltage due to the high value of capacitance in the circuit. Since the run winding is highly inductive, the current through it lags the applied voltage. Notice that the amount of phase separation now approaches 90°, or an actual two-phase situation. The starting torque produced by a capacitor-start induction motor is much greater than for the split-phase motor. Thus this type of motor can be used for applications requiring greater initial torque. However, they are somewhat more expensive than split-phase ac motors. Most capacitor motors, as well as split-phase motors, are used in fractional-horsepower sizes (less than 1 hp).

Another type of capacitor motor is called a capacitor-start, capacitor-run, or two-value capacitor motor. Its circuit is shown in Figure 9-10. This motor employs two capacitors. One, of higher value, is in series with the start winding and centrifugal switch. The other, of lower value, is in series with the start winding and remains in the

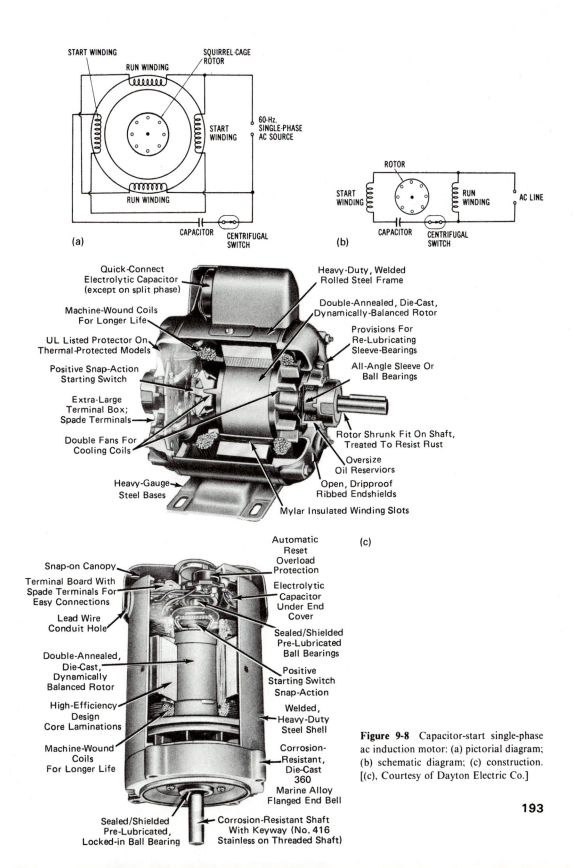

Figure 9-8 Capacitor-start single-phase ac induction motor: (a) pictorial diagram; (b) schematic diagram; (c) construction. [(c), Courtesy of Dayton Electric Co.]

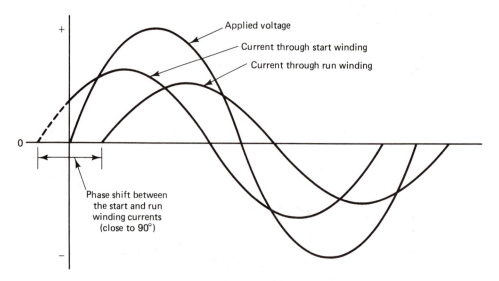

Figure 9-9   Relationship of voltage and current in a capacitor-start single-phase ac induction motor.

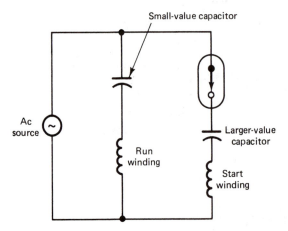

Figure 9-10   Schematic of a capacitor-start, capacitor-run ac induction motor.

circuit during operation. The larger capacitor is used only to increase starting torque, and is removed from the circuit by the centrifugal switch during normal operation. The smaller capacitor and the entire start winding are part of the operational circuit of the motor. The smaller capacitor helps to produce a more constant running torque, as well as quieter operation and improved power factor.

Yet another type of capacitor motor is called a permanent capacitor motor, shown in Figure 9-11. This motor has no centrifugal switch, so its capacitor is permanently connected into the circuit. These motors are used only for very low torque requirements and are made in small, fractional-horsepower sizes.

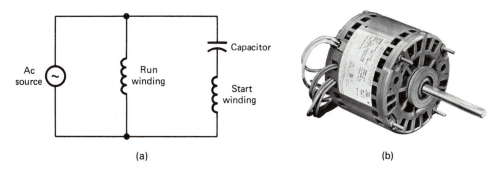

Figure 9-11 Permanent capacitor ac induction motor: (a) schematic diagram; (b) permanent capacitor blower motor. [(b), Courtesy of Marathon Electric.]

Both split-phase motors and capacitor motors may have their direction of rotation reversed easily. We simply change the relationship of the start winding and run winding. By reversing either the start winding connections or the run winding connections (but not both), the rotational direction will be reversed.

Shaded-pole induction motors. Another method of producing torque by a simulated two-phase method is called *pole shading*. Shaded-pole motors are used for very low torque applications such as fans and blower units (see Figure 9-12). They are low-cost, rugged, and reliable motors that are made only in low-horsepower ratings, usually from $\frac{1}{300}$ to $\frac{1}{3}$ hp.

The operational principle of a shaded-pole motor is shown in Figure 9-13. The single-phase alternation shown is for discussion purposes only. The dashed lines represent induced voltage into the shaded section of the field poles. Note the shading

Figure 9-12 Shaded-pole blower motor.

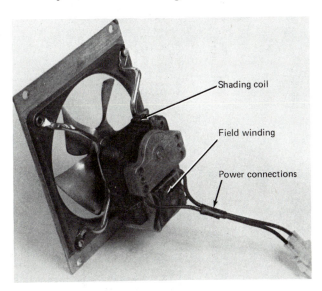

Shading coil

Field winding

Power connections

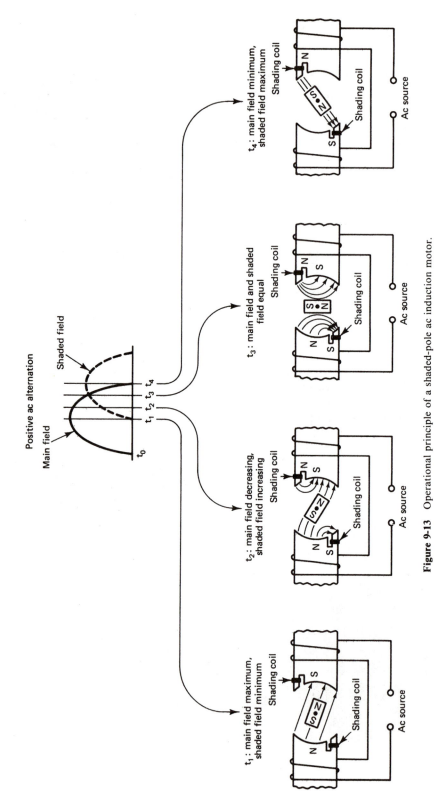

Figure 9-13 Operational principle of a shaded-pole ac induction motor.

coils in the upper right and lower left of the two poles. The shaded pole is encircled by a heavy copper conductor and is actually a part of the main field pole. This closed-loop conductor will have current induced into it when ac is applied to the field.

When ac voltage is applied to the stator windings, the magnetic flux in the main poles induces a voltage into the shaded sections of the poles. Since the shaded section acts like a transformer secondary, its voltage is out of phase with the main field voltage as shown on the waveform diagram. Note the four time intervals shown in sequence in Figure 9-13. The voltage induced in the shaded pole from the main pole field causes movement of the rotor to continue. Study this illustration to understand the operational principle of the shaded-pole ac induction motor.

The shaded-pole motor is inexpensive since it uses a squirrel-cage rotor and has no auxiliary starting winding or centrifugal mechanism. Applications are limited mainly to small fans and blowers.

Repulsion motors. Another type of ac induction motor is the repulsion motor. This motor was used for many applications, but is now being replaced by other types of single-phase motors. The principle of operation of the repulsion motor is an interesting contrast to other induction motors. Motors of this type should be studied in relation to their operational principles.

Figure 9-14 shows the operational principle of a repulsion motor. This motor has a wound rotor that functions similarly to a squirrel-cage rotor. It also has a commutator/brush assembly. The brushes are shorted together to produce an effect similar to the shorted conductors of a squirrel-cage rotor. The position of the brush

Figure 9-14 Operational principle of a repulsion motor.

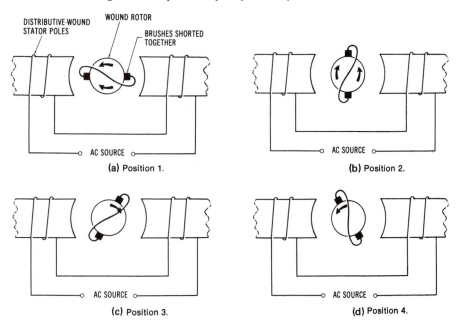

axis determines the amount of torque developed and the direction of rotation of the repulsion motor.

In position 1 of Figure 9-14 the brush axis is aligned horizontally with the stator poles. Equal and opposite currents are now induced into both halves of the rotor. Thus no torque is developed with the brushes in this position. In position 2, the brushes are placed at a 90° angle to the stator field poles. The voltages induced into the rotor again counteract one another and no torque is developed. In position 3, the brush axis is shifted about 60° from the stator poles. The current flow in the armature now causes a magnetic field to exist around the rotor. The rotor field will now follow the revolving stator field in a clockwise direction. As might be expected, when we shift the brush axis in the opposite direction, as shown in position 4, rotation reversal will result. Thus magnetic repulsion between the stator field and the induced rotor field causes the rotor to turn in the direction of the brush shift.

Repulsion-start induction motors and some similar types of modified repulsion motors have very high starting torque. Their speed can be varied by varying the position of the brush axis. However, the mechanical problems inherent with this type of motor have caused it to become obsolete.

SINGLE-PHASE SYNCHRONOUS MOTORS

It is often desirable, in timing or clock applications, to use a constant-speed drive motor. Such a motor, which operates from a single-phase ac line, is called a synchronous motor. The single-phase synchronous motor, such as the one shown in Figure 9-15, has stator windings that are connected across the ac line. The rotor,

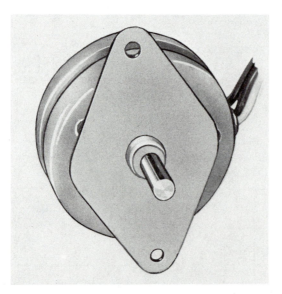

Figure 9-15 Single-phase synchronous motor. (Courtesy of Airpax Corp.)

however, is made of permanent-magnetic material. Some external method of starting must be used in order to initiate rotation. Once the rotor is started, it will rotate in synchronism with the revolving stator field, since it does not rely on the induction principle. The speeds of synchronous motors are based on the speed formula. For 60-Hz operation, the following synchronous speeds would be obtained:

Two-pole	3600 r/min
Four-pole	1800 r/min
Six-pole	1200 r/min
Eight-pole	900 r/min
Ten-pole	720 r/min
Twelve-pole	600 r/min

Single-phase synchronous motors are used only for very low torque applications. Such applications include clocks, phonograph drives, and timing devices requiring constant speed.

MOTOR DIMENSIONS

There are many different sizes of electric motors. *Fractional-horsepower* motors are those less than 1 hp in size, while integral-horsepower motors are larger than 1 hp. When specifying types of electric motors, one must know the dimensions of a motor to be used for a specific application. Dimensions are usually listed in a manufacturer's catalog. For several applications, motor method is also important. The major types of motor mountings are resilient, rigid, and flange methods. Resilient mountings provide a vibration-reducing insulated contact between the motor and the mounting surface. Rigid and flange mounting methods are used to attach a motor directly to a machine. These methods are illustrated in Figure 9-16. The dimensions of a typical motor are shown in Figure 9-17.

MOTOR NAMEPLATE DATA

It is important to be able to interpret the data on a motor nameplate. The information contained on a typical nameplate is summarized as follows:

Manufacturing company: the company that built the motor

Motor type: a specific type of motor (i.e., split-phase ac, universal, three-phase induction, etc.)

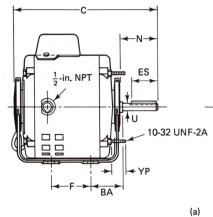
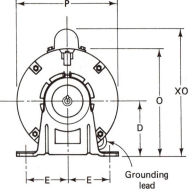

(a)

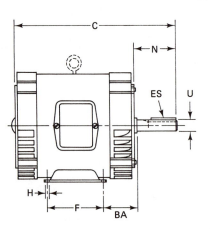
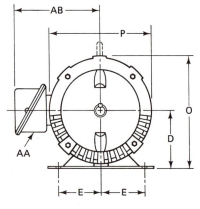

(b)

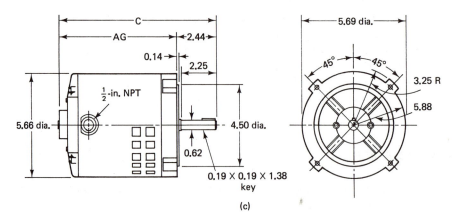

(c)

Figure 9-16 Types of motor frame mountings—side and end views with dimensions: (a) resilient mounting—capacitor-start motor; (b) rigid mounting—three-phase ac motor; (c) flange mounting—split-phase ac motor. (Courtesy of Marathon Electric.)

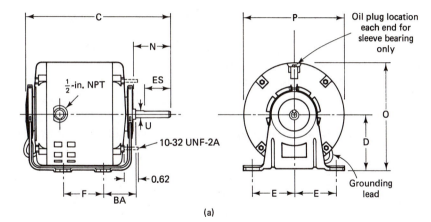

(a)

Dimensions[a] (in.)									
C	D	E	F	N	O	P	U	BA	ES
9.09	3.00	2.12	2.75	1.59	5.84	5.66	0.50	2.50	1.25 flat
9.34	3.00	2.12	2.75	1.59	5.84	5.66	0.50	2.50	1.25 flat
10.47	3.50	2.44	3.00	1.97	6.33	5.66	0.62	2.75	0.19 X 0.19 X 1.38 key
11.22	3.50	2.44	3.00	1.97	6.33	5.66	0.62	2.75	0.19 X 0.19 X 1.38 key
10.09	3.00	2.12	2.75	1.59	5.83	5.66	0.50	2.50	1.25 flat
10.47	3.50	2.44	3.00	1.97	6.62	6.25	0.62	2.75	0.19 X 0.19 X 1.38 key
9.47	3.50	2.44	3.00	1.97	6.33	5.66	0.50[b]	2.75	0.19 X 0.19 X 1.38 key
9.72	3.50	2.44	3.00	1.97	6.33	5.66	0.50[b]	2.75	0.19 X 0.19 X 1.38 key
10.47	3.50	2.44	3.00	1.97	6.33	5.66	0.50[b]	2.75	0.19 X 0.19 X 1.38 key
9.47	3.00	2.12	2.75	1.97	5.83	5.66	0.50	2.50	1.62 flat
9.72	3.00	2.12	2.75	1.97	5.83	5.66	0.50	2.50	1.62 flat
10.47	3.00	2.12	2.75	1.97	5.83	5.66	0.50	2.50	1.62 flat

Various motor types available

[a] Dimensions from side and end views.
[b] Motor is supplied with 0.62 adapter sleeve.

(b)

Figure 9-17 Typical ac motor: (a) resilient mounting—split-phase ac motor: side and end views with dimensions lettered; (b) sample dimensions for the motor. (Courtesy of Marathon Electric.)

Identification number: number assigned by the manufacturer

Model number: number assigned by the manufacturer

Frame type: frame size defined by NEMA

Number of phases (ac): single-phase or three-phase

Horsepower: the amount produced at rated speed

Cycles (ac): frequency the motor should be used with (usually 60 Hz)

Speed (r/min): the amount at rated horsepower, voltage, and frequency

Voltage rating: operating voltage of motor

Current rating (amperes): current drawn at rated load, voltage, and frequency

Thermal protection: indicates the type of overload protection used

Temperature rating (° C): amount of temperature that the motor will rise over ambient temperature, when operated

Time rating: time the motor can be operated without overheating (usually continuous)

Amperage: current drawn at rated load, voltage, and frequency

REVIEW

9.1. What are the three basic types of single-phase ac motors?

9.2. Discuss the construction and operation of a universal motor.

9.3. What are some applications of universal motors?

9.4. Discuss the operation of single-phase ac induction motors.

9.5. How is the speed of a single-phase ac induction motor determined?

9.6. What is meant by the term "slip"?

9.7. How is rotor frequency of a single-phase ac induction motor determined?

9.8. What are some common types of single-phase ac induction motors?

9.9. Discuss the construction and operation of a split-phase ac induction motor.

9.10. What is the purpose of a centrifugal switch in a single-phase ac induction motor?

9.11. What are the classes of induction motors according to starting characteristics?

9.12. Discuss the construction and operation of a capacitor-start ac induction motor.

9.13. What is a capacitor-start, capacitor-run motor?

9.14. What is a permanent capacitor motor?

9.15. Discuss the construction and operation of a shaded-pole ac induction motor.

9.16. Discuss the construction and operation of a repulsion motor.

9.17. Discuss the construction and operation of a single-phase synchronous motor.

9.18. How is the direction of rotation of each of the following single-phase motors reversed: **(a)** split-phase motor, **(b)** capacitor-start motor, **(c)** shaded-pole motor, **(d)** repulsion motor, and **(e)** synchronous motor?

9.19. What is a fractional-horsepower motor? An integral-horsepower motor?

9.20. What are the major types of motor mountings?

PROBLEMS

9.1. Calculate the synchronous speeds of the following single-phase ac motors.
   (a) Two-pole, 50 Hz   **(b)** Four-pole, 60 Hz
   (c) Eight-pole, 60 Hz   **(d)** Six-pole, 50 Hz

9.2. What is the percent slip of a single-phase ac induction motor with a synchronous speed of 3600 r/min and a rotor speed of 3450 r/min?

9.3. What is the rotor frequency of the motor of Problem 9.2?

9.4. What is the percent slip of a four-pole ac induction motor operating at a speed of 1725 r/min on a 60-Hz power line?

9.5. What is the rotor frequency of the motor of Problem 9.4?

9.6. What is the efficiency of a 240-V single-phase ac induction motor which is rated at 3 hp and converts 2650 W of true power?

9.7. What is the efficiency of a $\frac{1}{2}$-hp 120-V single-phase ac induction motor which converts 500 W of true power?

TEN

Three-Phase
Alternating-Current Motors

Three-phase ac motors are often called the "workhorses of industry." Most motors used in industry are operated from three-phase power sources. There are three basic types of three-phase motors: (1) three-phase induction motors, (2) three-phase synchronous motors, and (3) three-phase wound-rotor motors. Several types of three-phase motors are shown in Figure 10-1.

THREE-PHASE INDUCTION MOTORS

The three-phase induction motor is illustrated in Figure 10-2(a). A three-phase ac waveform is shown at the top of the illustration in Figure 10-2(b). These ac waveforms represent the input voltage to phases A, B, and C of the stator of a three-phase induction motor. At reference point $t1$ on the diagram phase A has a voltage value of zero, phase B is approaching maximum negative value, and phase C is approaching maximum positive value. The resulting polarities of the stator coils are shown in the diagram below point $t1$. Phase B has a north polarity at area B_2 and a south polarity at B_1. Phase C has a north polarity at area C_1 and a south polarity at C_2. Since the magnetic flux associated with each phase travels from north to south through the area of least reluctance, the resulting magnetic flux develops as shown in the diagram. The arrow represents the position of a rotor placed within the machine. The arrowhead represents the north polarity of the rotor.

At reference point $t2$ of the illustration, phase C has a voltage value of zero, so minimum magnetic field flux will develop around the coil of phase C. Phase A is

Power Connections
Squirrel-Cage Rotor
Laminated Stator
Stator Windings
Cooling Fins
Bearing

(a)

SIEMENS-ALLIS

(b)

(c)

Figure 10-1 (a) Cutaway of a three-phase induction motor; (b) large three-phase induction motor; (c) cutaway of a small three-phase ac motor; (d) cutaway of a three-phase ac motor with parts labeled; (e) cutaway of a high-efficiency three-phase induction motor; (f) large integral-horsepower three-phase ac motor. [(a), Courtesy of Marathon Electric; (b), (e), courtesy of Siemens-Allis, Inc.; (c), courtesy of Century Electric, Inc.; (d), courtesy of General Electric Co., Industrial Motor Div.; (f), courtesy of Lima Electric Co., Inc.]

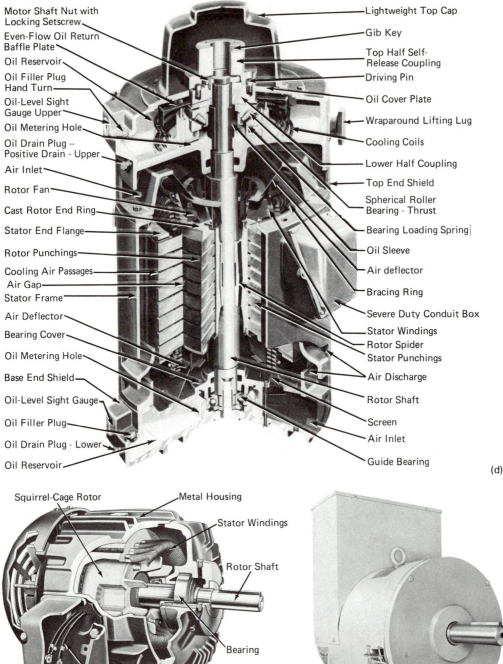

Motor Shaft Nut with Locking Setscrew
Even-Flow Oil Return Baffle Plate
Oil Reservoir
Oil Filler Plug Hand Turn
Oil-Level Sight Gauge Upper
Oil Metering Hole
Oil Drain Plug -- Positive Drain - Upper
Air Inlet
Rotor Fan
Cast Rotor End Ring
Stator End Flange
Rotor Punchings
Cooling Air Passages
Air Gap
Stator Frame
Air Deflector
Bearing Cover
Oil Metering Hole
Base End Shield
Oil-Level Sight Gauge
Oil Filler Plug
Oil Drain Plug - Lower
Oil Reservoir

Lightweight Top Cap
Gib Key
Top Half Self-Release Coupling
Driving Pin
Oil Cover Plate
Wraparound Lifting Lug
Cooling Coils
Lower Half Coupling
Top End Shield
Spherical Roller Bearing - Thrust
Bearing Loading Spring
Oil Sleeve
Air deflector
Bracing Ring
Severe Duty Conduit Box
Stator Windings
Rotor Spider
Stator Punchings
Air Discharge
Rotor Shaft
Screen
Air Inlet
Guide Bearing

(d)

Squirrel-Cage Rotor
Metal Housing
Stator Windings
Rotor Shaft
Bearing
Three-Phase AC Power Lines

(e)

(f)

Figure 10-1 (*cont.*)

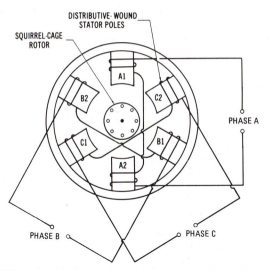

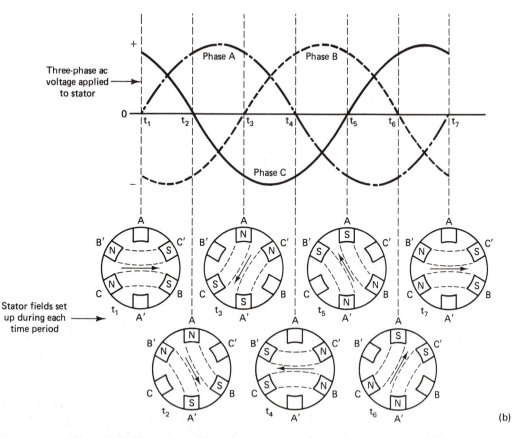

Figure 10-2 Three-phase ac induction motor operational principle: (a) pictorial diagram; (b) three-phase waveforms and the resulting stator fields; (c) wye and delta connections.

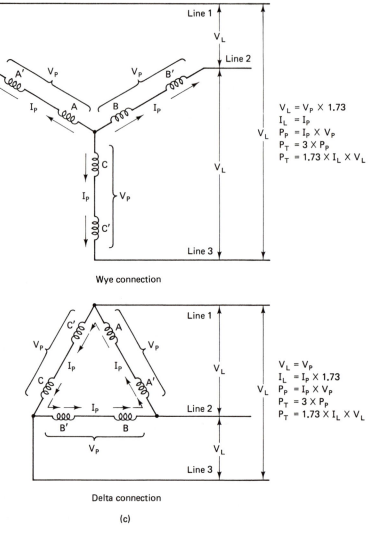

$$V_L = V_P \times 1.73$$
$$I_L = I_P$$
$$P_P = I_P \times V_P$$
$$P_T = 3 \times P_P$$
$$P_T = 1.73 \times I_L \times V_L$$

Wye connection

$$V_L = V_P$$
$$I_L = I_P \times 1.73$$
$$P_P = I_P \times V_P$$
$$P_T = 3 \times P_P$$
$$P_T = 1.73 \times I_L \times V_L$$

Delta connection

(c)

Figure 10-2 (*cont.*)

approaching maximum positive value and phase B is approaching maximum negative value. Phase A has a north polarity at A_1 and a south polarity at A_2. Phase B has a south polarity at B_1 and a north polarity at B_2. The resulting magnetic flux of the stator causes the rotor to shift a total of 60°, which is the amount of physical spacing between coils.

At reference point $t3$, phase B is at minimum value, phase A is approaching maximum positive and phase C is approaching maximum negative. Phase B has minimum magnetic flux. Phase A has a north polarity at A_1 and a south polarity at A_2. Phase C has a north polarity at C_2 and a south polarity at C_1. The resulting magnetic flux causes the rotor to shift another 60°.

At reference point *t*4, the sequence begins with opposite polarities developing around the stator coils. Notice that the polarities of B_1, B_2, C_1, and C_2 are opposite from those of reference point *t*1. The resulting polarities cause the rotor to position itself 180° from the position of reference point *t*1.

The sequence continues as the three-phase induction motor continues to operate. The actual rotor of this type of motor is a squirrel-cage type that relies on the induction principle to sustain operation. The application of three-phase voltage to the motor provides phase separation which allows this type of motor to be self-starting.

Figure 10-2(c) shows the voltage, current, and power relationships of the stator coils of a three-phase induction motor. These are the same for any three-phase system. Notice that voltage across each phase in the wye connection is less than in the delta connection, since in the wye connection $V_P = V_L \div 1.73$. This is important for motor starting applications, which are discussed in Chapter 12.

The construction of this motor is very simple. It has only a distributive-wound stator, which is connected in either a wye or a delta configuration, and a squirrel-cage rotor. Since three-phase voltage is applied to the stator, phase separation is already established. No external starting mechanisms are needed. Three-phase induction motors are made in a variety of integral-horsepower sizes and have good starting and running torque characteristics.

The direction of rotation of a three-phase motor of any type can be changed very easily. If any two power lines coming into the stator windings are reversed, the direction of rotation of the shaft will change. Three-phase induction motors are used for many industrial applications, such as mechanical energy sources for machine tools, pumps, elevators, hoists, conveyors, and many other industrial systems.

THREE-PHASE SYNCHRONOUS MOTORS

Three-phase synchronous motors are unique and very specialized motors. They are considered a constant-speed motor and they can be used to correct the power factor of three-phase systems. Figure 10-3 shows a pictorial diagram of a three-phase ac synchronous motor.

The three-phase synchronous motor is physically constructed like a three-phase alternator. Direct current is applied to the rotor to produce a rotating electromagnetic field, and the stator windings are connected in either a wye or a delta configuration. The only difference is that three-phase ac power is applied to the synchronous motor, while three-phase power is extracted from the alternator. Thus the motor acts as an electrical load, while the alternator functions as a source of three-phase power. This relationship should be kept in mind during the following discussion.

The three-phase synchronous motor differs from the three-phase induction motor in that the rotor is wound and is connected through a slip-ring/brush assembly to a dc power source. Three-phase synchronous motors, in their pure form, have no starting torque. Some external means must be used to start the motor initially. Synchronous motors are constructed so that they will rotate at the same speed as the

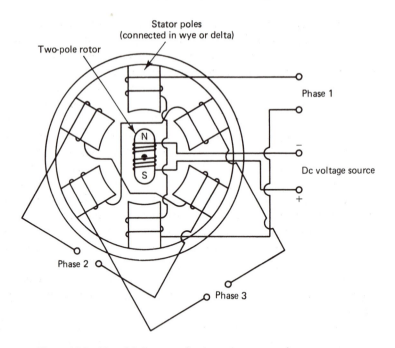

Figure 10-3 Pictorial diagram of a three-phase ac synchronous motor.

revolving stator field. At synchronous speed, rotor speed equals stator speed and the synchronous motor has zero slip. Thus the speed of a synchronous motor may be determined by using the following formula:

$$S = \frac{f \times 120}{N/3}$$

where S = speed of a synchronous motor in r/min
$\quad\quad f$ = frequency of the applied ac voltage in hertz
$\quad N/3$ = number of stator poles per phase
$\quad\quad 120$ = a conversion constant

Note that this is the same as the formula used to determine the stator speed of a single-phase motor except that the number of poles must be divided by 3 (the number of phases). A three-phase motor with 12 actual poles would have four poles per phase. Therefore, its stator speed would be 1800 r/min. Synchronous motors have operating speeds that are based on the number of stator poles they have.

Synchronous motors are usually employed in very large horsepower ratings. One method of starting a large synchronous motor is to use a smaller auxiliary dc machine connected to the shaft of the synchronous motor, as illustrated in Figure 10-4. The method of starting would be as follows:

Step 1: Dc power is applied to the auxiliary motor, causing it to increase in speed. Three-phase ac power is applied to the stator.

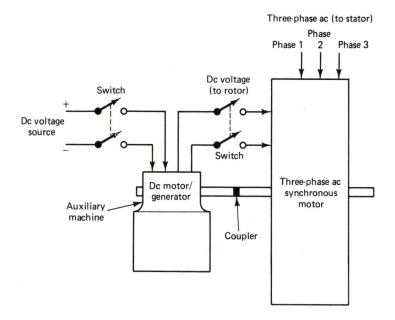

Figure 10-4 Auxiliary motor used to start a three-phase synchronous motor.

Step 2: When the speed of rotation reaches a value near the synchronous speed of the motor, the dc power circuit is opened and, at the same time, the terminals of the auxiliary machine are connected across the commutator/brush assembly of the rotor.

Step 3: The auxiliary machine now converts to generator operation and supplies exciter current to the rotor of the synchronous motor, using the motor as its prime mover.

Step 4: Once the rotor is magnetized, it will "lock" in step or synchronize with the revolving stator field.

Step 5: The speed of rotation will remain constant under changes in load condition.

Another starting method is shown in Figure 10-5. This method utilizes "damper" windings, which are similar to the conductors of a squirrel-cage rotor. These windings are placed within the laminated iron of the rotor assembly. No auxiliary machine is required when damper windings are used. The starting method used is as follows:

Step 1: Three-phase ac power is applied to the stator windings.

Step 2: The motor will operate as an induction motor due to the "transformer action" of the damper windings.

Step 3: The motor speed will build up so that the rotor speed is somewhat less than the speed of the revolving stator field.

Step 4: Dc power from a rotating dc machine or, more commonly, a

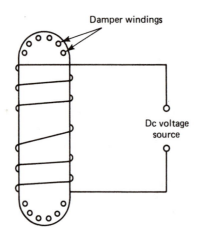

Damper windings

Dc voltage source

Figure 10-5   Damper windings used in the rotor of a three-phase synchronous motor.

rectification system is applied to the commutator/brush assembly of the rotor.

Step 5: The rotor becomes magnetized and builds up speed until rotor speed is equal to stator speed.

Step 6: The speed of rotation remains constant regardless of the load placed on the shaft of the motor.

An outstanding advantage of the three-phase synchronous motor is that it can be connected to a three-phase power system to increase the overall power factor of the system.

POWER FACTOR IN AC SYSTEMS

In dc circuits, power is equal to the product of voltage and current ($P = V \times I$). This formula is true also for purely resistive circuits. However, when reactance (either inductive or capacitive) is present in an ac circuit, power is no longer a product of voltage and current. Reactive circuits cause changes in the method used to compute power in ac circuits.

The product of voltage and current in ac circuits is expressed in voltamperes (VA) or kilovoltamperes (kVA) and is known as *apparent power*. When using meters to measure power in an ac circuit, apparent power is the voltage reading multiplied by the current reading. The actual power that is converted to another form of energy by the circuit is measured with a wattmeter. This actual power is referred to as true power. The ratio of true power converted in a circuit to apparent power is called the *power factor* and is expressed as

$$\text{PF} = \frac{P}{\text{VA}}$$

or

$$\% \text{ PF} = \frac{P}{\text{VA}} \times 100$$

where PF = power factor of the circuit
          P = true power in watts
         VA = apparent power in voltamperes

The maximum value of power factor is 1.0, or 100%, which would be obtained in a purely resistive circuit. This is referred to as *unity power factor.*

The phase angle between voltage and current in an ac circuit determines the power factor. If a purely inductive or capacitive circuit existed, the 90° phase angle would cause a power factor of zero to result. In practical ac circuits, the power factor varies according to the relative values of resistance and reactance.

The power relationships are simplified by using the power triangle discussed in Chapter 4. There are two components that affect the power relationship in an ac circuit. The in-phase (resistive) component which results in power conversion in the circuit is called *active power.* Active power is the true power of the circuit and is measured in watts. The second component is that which results from an inductive or capacitive reactance and is 90° out of phase with the active power. This component, called *reactive power,* does not produce an energy conversion in the circuit. Reactive power is measured in voltamperes reactive (var).

The cosine of the phase angle of the power triangle is expressed as

$$\text{cosine } \theta = \frac{\text{adjacent side}}{\text{hypotenuse}} \quad \text{or} \quad \text{cosine } \theta = \frac{\text{true power (W)}}{\text{apparent power (VA)}}$$

Therefore, true power can be expressed as $\text{W} = \text{VA} \times \text{cosine } \theta$. Note that the expression true power / apparent power is the power factor of a circuit. Therefore, the power factor is equal to the cosine of the phase angle (PF = cosine θ).

POWER-FACTOR CORRECTION

Since most industries use a large number of electric motors, the industrial plant represents a highly inductive load. This means that industrial power systems operate at a power factor of less than unity (1.0). It is undesirable for an industry to operate at a low power factor, since the utility company would have to supply more electrical power to the industry than is actually used.

A given value of voltamperes (voltage × current) is supplied to an industry. If the power factor of the industry is low, the current must be higher since the power converted by the total industrial load equals VA × PF. The value of power factor decreases as the reactive power drawn by the industry increases. This is illustrated in Figure 10-6. Assume a constant value of true power to see the effect of increases in reactive power drawn by an inductive load. The smallest reactive power shown (var_1)

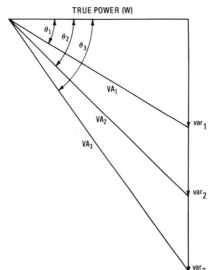

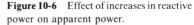

Figure 10-6 Effect of increases in reactive power on apparent power.

results in the voltampere value of VA_1. As reactive power is increased, as shown by the var_2 and var_3 values, more voltamperes (VA_2 and VA_3) must be drawn from the source. This is true since the voltage component of the supplied voltamperes remains constant. This example represents the same effect as a decrease in power factor, since PF = W/VA, and as VA increases, PF will decrease if W remains constant.

Utility companies usually charge industries for operating at power factors below a specified level. It is desirable for industries to "correct" their power factor to avoid such charges and to make more economical use of electrical energy. Two methods can be used to increase the power factor: (1) power factor-corrective capacitors and (2) three-phase synchronous motors. Since the effect of capacitive reactance is opposite to that of inductive reactance, the reactive effects will counteract one another. Either power-factor-corrective capacitors or three-phase synchronous motors can be used to add the effect of capacitance to an ac power line.

In the example shown in Figure 10-7, assume that both true power and inductive reactive power remain constant at values of 10 kW and 10 kvar. In Figure 10-7(a), the power factor equals 70%. If 5-kvar capacitive reactive power is introduced into the ac power line, the net reactive power becomes 5 kvar (10-kvar inductive minus 5-kvar capacitive), as shown in Figure 10-7(b). With the addition of 5-kvar capacitive, the power factor is increased to 89%. Now, in Figure 10-7(c), if 10-kvar capacitive is added to the ac power line, the total reactive power becomes zero. The true power is now equal to the apparent power; therefore, the power factor is 1.0, or 100%, which is characteristic of a purely resistive circuit. The effect of the increased capacitive reactive power is to increase or "correct" the power factor and thus reduce the current drawn from the ac power lines that supply the industrial loads. It may be beneficial for industries to invest in either power factor-corrective capacitors or three-phase synchronous motors to correct their power factor.

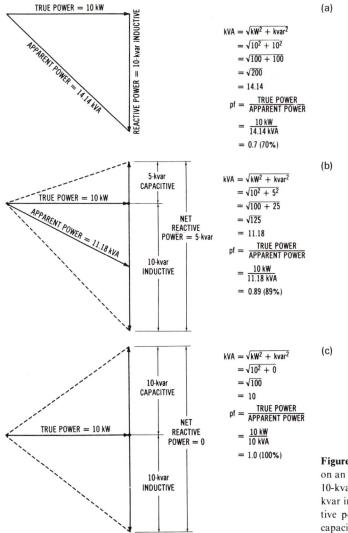

$$kVA = \sqrt{kW^2 + kvar^2}$$
$$= \sqrt{10^2 + 10^2}$$
$$= \sqrt{100 + 100}$$
$$= \sqrt{200}$$
$$= 14.14$$

$$pf = \frac{TRUE\ POWER}{APPARENT\ POWER}$$
$$= \frac{10\ kW}{14.14\ kVA}$$
$$= 0.7\ (70\%)$$

(a)

$$kVA = \sqrt{kW^2 + kvar^2}$$
$$= \sqrt{10^2 + 5^2}$$
$$= \sqrt{100 + 25}$$
$$= \sqrt{125}$$
$$= 11.18$$

$$pf = \frac{TRUE\ POWER}{APPARENT\ POWER}$$
$$= \frac{10\ kW}{11.18\ kVA}$$
$$= 0.89\ (89\%)$$

(b)

$$kVA = \sqrt{kW^2 + kvar^2}$$
$$= \sqrt{10^2 + 0}$$
$$= \sqrt{100}$$
$$= 10$$

$$pf = \frac{TRUE\ POWER}{APPARENT\ POWER}$$
$$= \frac{10\ kW}{10\ kVA}$$
$$= 1.0\ (100\%)$$

(c)

Figure 10-7 Effect of capacitive reactance on an inductive load: (a) reactive power = 10-kvar inductive; (b) reactive power = 10-kvar inductive, 5-kvar capacitive; (c) reactive power = 10-kvar inductive, 10-kvar capacitive.

Utility companies also attempt to correct the power factor of their systems. A certain quantity of inductance is present in most of the power distribution system, including the generator windings, the transformer windings, and the power lines. To counteract the inductive effects, utility companies use power factor-corrective capacitors, such as the pole-mounted units shown in Figure 10-8. This type of capacitor unit, plus larger banks of capacitors located at substations, can be used to correct the power factor of ac power systems.

Three-phase synchronous motors are sometimes used only to correct system power factor. If no load is intended to be connected to the shaft of a three-phase synchronous motor, it is called a *synchronous capacitor*. It is designed to act only as a

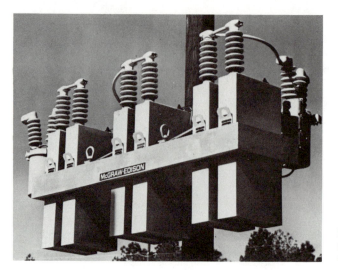

Figure 10-8 Pole-mounted power-factor-correction capacitors. (Courtesy of Mc-Graw-Edison Co., Power Systems Group.)

power-factor-corrective machine. Of course, it might be beneficial to use this motor as a constant-speed drive connected to a load as well as for power-factor correction.

We know from previous discussion that low power factors cannot be tolerated by industries. The expense of installing three-phase synchronous machines could be justified in order to appreciably increase system power factor. To understand how a three-phase synchronous machine operates as a power factor-corrective machine, refer to the curves of Figure 10-9. Synchronous motors operate at a constant speed; thus variation in rotor dc excitation current has no effect on speed. The excitation level will change the power factor at which the machine operates. Three operational conditions may exist, depending on the amount of dc excitation applied to the rotor.

Figure 10-9 Operating characteristic curves of a three-phase synchronous motor: (a) rotor current versus stator current; (b) rotor current versus power factor.

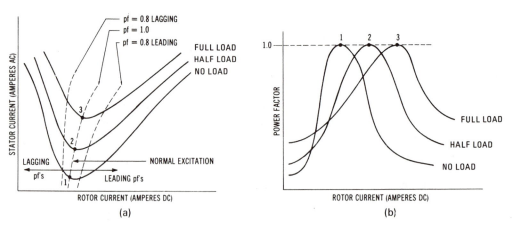

These conditions are:

1. *Normal excitation:* operates at a power factor of 1.0
2. *Underexcitation:* operates at a lagging power factor (inductive effect)
3. *Overexcitation:* operates at a leading power factor (capacitive effect)

Note the variation of stator current drawn by the synchronous motor as rotor current varies. Observe also that stator current is minimum when power factor equals 1.0 or 100%. The situations shown on the graph indicate no-load, half-load, and full-load conditions with power factors equal to 1.0, 0.8 leading, and 0.8 lagging. These curves are sometimes referred to as V-curves for a synchronous machine. The other graph shows the variation of power factor with changes in rotor current under three different load conditions. Thus a three-phase synchronous motor, when overexcited, can improve the overall power factor of a three-phase system.

As the load increases, the angle between the stator pole and the corresponding rotor pole on the synchronous machine increases. The stator current will also increase. However, the motor will remain synchronized unless the load causes "pull-out" to take place. The motor would then stop rotating due to the excessive torque required to rotate the load. Most synchronous motors are rated larger than 100 hp and are used for many industrial applications requiring constant-speed drives.

WOUND-ROTOR INDUCTION MOTORS

The wound-rotor induction motor, shown in Figure 10-10, is a specialized type of three-phase motor. This motor may be controlled externally by placing resistances in series with its rotor circuit. The starting torque of a wound-motor induction motor can be varied by the value of external resistance. The advantages of this type of motor are a lower starting current, a high starting torque, smooth acceleration, and ease of control.

Figure 10-10   Three-phase wound-rotor induction motor.

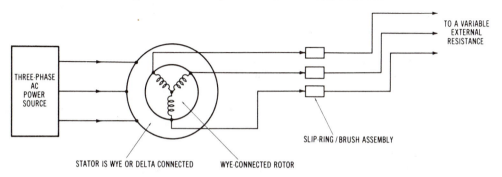

The major disadvantage of this type of motor is that it costs a great deal more than an equivalent three-phase induction motor using a squirrel-cage rotor. Thus they are not used as extensively as other three-phase motors.

REVIEW

10.1. What are the three basic types of three-phase motors?

10.2. Discuss the construction and operation of a three-phase induction motor.

10.3. How can the direction of rotation of a three-phase induction motor be reversed?

10.4. Discuss the construction and operation of a three-phase synchronous motor.

10.5. Discuss the two methods of starting three-phase synchronous motors.

10.6. What is meant by power-factor correction?

10.7. Why is power-factor correction important for industries?

10.8. What is meant by the following terms: **(a)** normal excitation, **(b)** underexcitation, and **(c)** overexcitation?

10.9. What is the relationship of rotor current and stator current in a three-phase synchronous motor at unity power factor?

10.10. What is the relationship between rotor current and power factor of a three-phase synchronous motor at full-load?

10.11. Discuss the construction and operation of a three-phase wound-rotor induction motor.

10.12. What are some advantages of wound-rotor induction motors?

PROBLEMS

10.1. Calculate the synchronous speeds of the following three-phase motors.
   (a) 6-pole, 60 Hz   **(b)** 12-pole, 60 Hz
   (c) 18-pole, 60 Hz   **(d)** 24-pole, 60 Hz

10.2. A 2000-kVA power system operates at a power factor of 0.6 lagging. Calculate the following:
   (a) True power of the system
   (b) Total reactive power—kvar
   (c) kvar required to obtain unity power factor
   (d) kvar required to obtain 0.8 power factor
   (e) Capacitance required to obtain unity power factor

10.3. An industry has an average load of 4500 kVA at a power factor of 0.6 lagging. An 800-hp three-phase synchronous motor with an efficiency of 80% is added to improve system power factor to 0.9 lagging. Calculate:
   (a) Power factor of the synchronous motor **(b)** kVA rating of the synchronous motor

10.4. A three-phase synchronous motor rated at a 0.8 power factor (leading) is connected across a 240-V three-phase power line. The power factor of the system prior to connection of the synchronous machine was 0.7 (lagging). What effect (increase or decrease) will the addition of the synchronous capacitor have on the following factors of the three-phase system:

(a) Power factor   **(b)** Line current
(c) Apparent power   **(d)** Reactive power?

Discuss why each factor changes as it does from a mathematical point of view.

10.5. When the mechanical load applied to the shaft of a three-phase induction motor increases, what effect on the following operational factors can be expected:

(a) Speed   **(b)** Line current
(c) Power factor   **(d)** Efficiency
(e) Torque   **(f)** Horsepower?

Discuss why each factor changes as it does from a mathematical point of view.

ELEVEN

Specialized Electrical Machines

Within the last ten years, some new and unique types of electrical machines have been developed. Particular emphasis is now being placed on the design of machines controlled by computer systems and used for automated manufacturing systems. New developments in industry have led to the use of several specialized electrical machines.

SYNCHRO AND SERVO SYSTEMS

In industry today there is a need for machines that produce a type of rotary motion that is somewhat different from that produced by electric motors. One type of machine, called a synchro system, employs rotary motion to control the angular position of one shaft by positioning the shaft of a second device. Synchro systems and servo-mechanisms are used in industry to achieve this operation. Through these devices, it becomes possible to transmit rotary motion between locations without direct mechanical linkage.

Synchro systems are classified as two or more motor-generator units connected together to permit the transmission of angular shaft positions by electromagnetic field changes. When an operator turns the generator shaft of the generator unit to a certain position, it automatically rotates the motor shaft to an equivalent position at a remote location. Through this type of system, it is possible to achieve accurate control over great distances. Computers often employ these units to determine physical changes that take place.

In synchro systems that require increased rotational torque or precise movements of a control device, *servomechanisms* are employed. A servomechanism is ordinarily a special type of ac or dc motor that drives a precision piece of equipment in specific increments. Systems that include servomechanisms generally require amplifiers and error-detecting devices to control angular displacement.

Synchro system operation.   A synchro system contains two or more electromagnetic devices that are similar in appearance to small electric motors. These devices are connected together in such a way that the angular position of the generator shaft can easily be transmitted to the motor or receiver unit. Figure 11-1 shows a schematic diagram and symbols of a synchro unit. As a general rule, the generator and motor units are identical electrically. Physically, the motor unit has a metal flywheel attached to its shaft to prevent shaft oscillations or vibrations when it is powered. The letters G and M inside the electrical symbol denote the generator or motor functions.

Figure 11-2 shows the circuit diagram of a basic synchro system. Single-phase ac

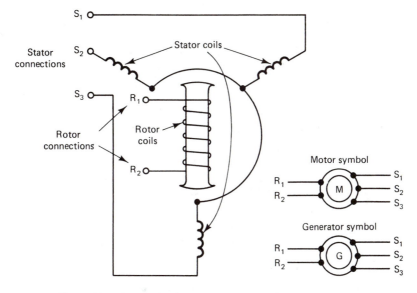

Figure 11-1   Schematic diagram and symbols of a synchro system.

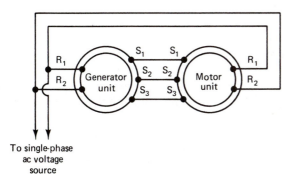

Figure 11-2   Circuit diagram of a synchro system.

voltage is used to power this particular system. The line voltage is applied to the rotors of both the generator and motor. The stationary coils, or stator windings, are connected together as indicated.

When power is initially applied to the system, the motor positions itself according to the location of the generator shaft. No physical change takes place after the motor unit aligns itself with the generator position. Both units remain in a stationary condition until some further action takes place. Turning the generator shaft a certain number of degrees in a clockwise direction causes a corresponding change in the motor unit. If calibrated dials are attached to the shaft of each unit, they would show the same angular displacement.

The stator coils of a synchro are wound inside a cylindrical laminated metal housing. The coils are uniformly placed in slots and connected to provide three poles spaced 120° apart. These coils serve functionally like the secondary windings of a transformer.

The rotor coil of the synchro unit is also wound on a laminated core. This type of construction causes north and south magnetic poles to extend from the laminated area of the rotor. Insulated slip rings on the shaft are used to supply ac power to the rotor. The rotor coil responds functionally like the primary windings of a transformer.

When ac is applied to the rotor coil of the synchro unit, it produces an alternating magnetic field. By transformer action, this field cuts across the stator coils and induces a voltage in each winding. The physical position of the rotor coil determines the amount of voltage induced in each stator coil. If the rotor coil is parallel with a stator coil, maximum voltage is induced. The induced voltage is minimum when the rotor coil is at right angles to a stator coil set.

The stator coils of the generator and motor of a synchro system are connected as indicated in the circuit diagram of Figure 11-2. Voltage induced in the stator coils of the generator therefore causes a resulting current flow in the stator coils of the motor. This in turn causes a corresponding magnetic field to be established in the motor stator. Line voltage applied to the rotor of the motor unit wall causes it to align itself with the magnetic field of the stator coils.

Any change in rotor position of the generator unit is translated into an induced voltage and applied to the motor stator coils. Through this action, linear displacement changes can be effectively transmitted to the motor through three rather small stator coil wires. Systems of this type are becoming very important in industrial automatic process control applications.

Servo systems.   Servo systems are rotating machines that accomplish such functions as changing the mechanical position or speed of a machine. Mechanical position applications include computer-controlled machinery and process control equipment. Speed applications are found in conveyor-belt control units, spindle speed control in machine tool operations, and disk or magnetic-tape drives for computers. A servo system is usually a rather complex unit that follows the commands of a closed-loop control path. Figure 11-3 shows the components of a typical servo system.

The input of a servo system serves as the reference element to which the

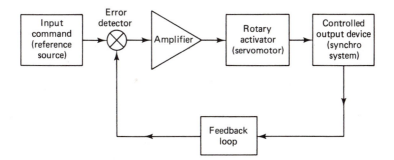

Figure 11-3   Block diagram of a servo system.

controlled device responds. By changing the input, a command is applied to the error detector. This device receives data from both the input source and from the controlled output device. If a correction is needed with reference to the input command, it is amplified and applied to the actuator. The actuator is normally a servomotor that produces controlled shaft displacements. The controlled output device is sometimes a synchro system that relays information back to the error detector for position comparison.

SERVOMOTORS

The servomotor is primarily responsible for producing mechanical changes from an electromagnetic actuating device. A device of this type is normally coupled to the controlled device by a gear train or some mechanical linkage. Both ac and dc servomotors, such as those shown in Figure 11-4, are used to achieve this operation.

Figure 11-4   Servomotors: (a) ac; (b) dc. [(a), Courtesy of Bodine Electric Co.; (b), courtesy of H.K. Porter Co., Inc.]

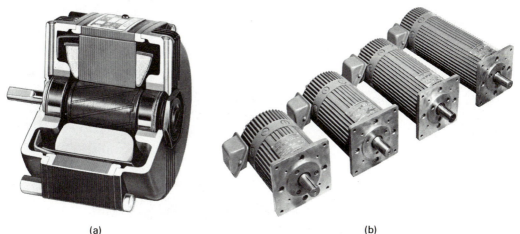

(a)   (b)

A servomotor is a very special type of device that is used to achieve a precise degree of rotary motion. Servomotors are unique when compared with other electric motors. Servomotors, for example, are designed to do something other than change electrical energy into rotating mechanical energy. Motors of this type must be able to respond accurately to signals developed by the systems amplifier. Second, they must be capable of reversing direction quickly when a specific signal polarity is applied. Also, the amount of torque developed by a servomotor must be quite high. As a general rule, torque is a function of the voltage and current levels of the supply source.

Two distinct types of servomotors are used today to achieve these operating conditions. An ac type of motor, called a *synchronous motor,* is commonly used in low-power applications. Excessive amounts of heat developed during starting conditions normally limit this motor to rather low output power applications. *Dc stepping motors* represent another type of servomotor.

AC Synchronous Motors

The construction of an ac synchronous motor is quite simple. It contains no brushes, commutators, slip rings, or centrifugal switches. As shown in Figure 11-5, it is simply made up of a rotor and a stator assembly. There is no direct physical contact between the rotor and stator. A carefully maintained air gap is always present between these two components. As a result of this construction, the motor has a long operating life and is highly reliable.

The speed of a synchronous motor is directly proportional to the frequency of the applied ac and the number of stator poles. Since the number of stator poles cannot be

Figure 11-5 Exploded view of an ac synchronous motor. (Courtesy of Superior Electric Co.)

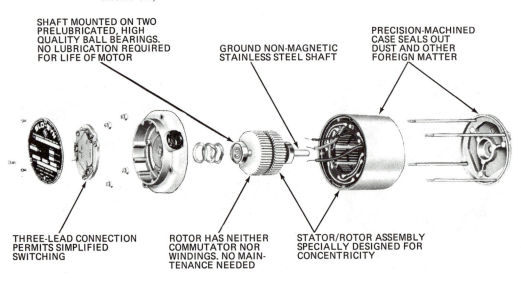

SHAFT MOUNTED ON TWO
PRELUBRICATED, HIGH
QUALITY BALL BEARINGS.
NO LUBRICATION REQUIRED
FOR LIFE OF MOTOR

GROUND NON-MAGNETIC
STAINLESS STEEL SHAFT

PRECISION-MACHINED
CASE SEALS OUT
DUST AND OTHER
FOREIGN MATTER

THREE-LEAD CONNECTION
PERMITS SIMPLIFIED
SWITCHING

ROTOR HAS NEITHER
COMMUTATOR NOR
WINDINGS. NO MAIN-
TENANCE NEEDED

STATOR/ROTOR ASSEMBLY
SPECIALLY DESIGNED FOR
CONCENTRICITY

effectively altered after the motor has been manufactured, frequency is the most significant speed factor. Speeds of 28, 72, and 200 r/min are typical, with 72 r/min being a common industrial standard.

Figure 11-6 shows the stator layout of a two-phase synchronous motor with four poles per phase. Poles N1-S3 and N5-S7 represent one phase while poles N2-S4 and N6-S8 represent the second phase. There are places for 48 teeth around the inside of the stator. One tooth per pole, however, has been eliminated to provide a space for the windings. Five teeth per pole, or a total of 40 teeth, are formed on the stator. The four coils of each phase are connected in series to achieve the correct polarity.

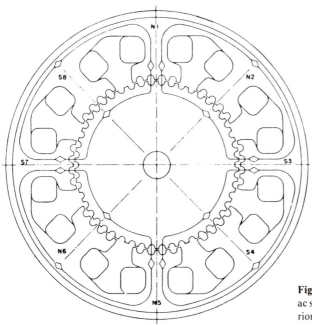

Figure 11-6 Stator layout of a two-phase ac synchronous motor. (Courtesy of Superior Electric Co.)

The rotor of the synchronous motor in Figure 11-5 is an axially magnetized permanent magnet. There are 50 teeth cast into its form. The front section of the rotor has one polarity while the back section has the opposite polarity. The physical difference in the number of stator teeth (40) and rotor teeth (50) means that only two teeth of each part can be properly aligned at the same time. With one section of the rotor being a north pole and the other section being a south pole, the rotor has the ability to stop very quickly. The rotor can also accomplish direction reversals very rapidly because of this gearlike construction.

A circuit diagram of a single-phase synchronous motor is shown in Figure 11-7. The resistor and capacitor of this circuit are used to produce a 90° phase shift in one winding. As a result, the two windings are always out of phase regardless of whether the switch is in the clockwise (CW) or counterclockwise (CCW) position. When power is applied, the four coils of one phase produce an electromagnetic field. The rotor is

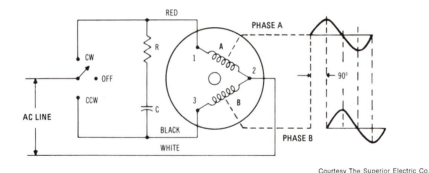

Figure 11-7 Circuit diagram of a single-phase synchronous motor. (Courtesy of Superior Electric Co.)

attracted to and aligns itself with these stator coils. Then, 90° later, the four coils of the second phase produce a corresponding field. The stator is again attracted to this position. As a result of this action, the rotor sees a moving force across first one phase and then the other. This force gives the rotor the needed torque that causes it to start and continue rotation when power is applied.

The synchronous motor just described has the capability of starting in one and one-half cycles of the applied ac source frequency. In addition to this, it can be stopped in five mechanical degrees of rotation. These two characteristics are attributed primarily to the geared rotor and stator construction. Synchronous motors of this type have another important characteristic. They draw the same amount of line current when stalled as they do when operating. This characteristic is very important in automatic machine-tool applications where overloads occur frequently.

DC Stepping Motors

Dc stepping motors are specialized electrical machines that are used to control automatic industrial equipment. Dc motors of this type are used in many high-power servomechanisms. They are typically more efficient and develop significantly more torque than the synchronous servomotor. The dc stepping motor is used primarily to change electrical pulses into rotary motion that is used to produce mechanical movement.

The shaft of a dc stepping motor rotates a specific number of mechanical degrees with each pulse of electrical input voltage. The amount of rotary movement or angular displacement produced by each pulse can be repeated precisely with each succeeding pulse from the input source. The resulting output of this machine is used to accurately locate or position automatic industrial systems.

The velocity, distance, and direction of a specific machine can be moved or controlled by a dc stepping motor. The movement error of this device is generally less than 5% per step. This error factor is not cumulative regardless of the distance moved or the number of steps taken. Motors of this type are energized by a dc drive amplifier that is controlled by computer logic circuits. The drive amplifier circuitry is a key factor in the overall performance of this motor.

The basic construction of a dc stepping motor is very similar to that of the ac synchronous motor of Figure 11-5. The stator construction and coil placement are also the same as that of the ac synchronous motor layout shown in Figure 11-6. Some manufacturers make servomotors that can be operated as either an ac synchronous motor or as a dc stepping motor. To achieve this, the rotor must be of permanent-magnet construction.

Dc stepping motor operation. An important principle that applies to the operation of dc stepping motors is the basic law of magnetism. Like magnetic poles repel and unlike magnetic poles attract. If a permanent-magnet rotor is placed between two series-connected stator coils, the situation of Figure 11-8 would result. With power applied to the stator, the rotor would be repelled in either direction. The direction of rotation in this case is unpredictable.

Adding two more stator coils to this simple motor, as indicated in Figure 11-9, would make the direction of rotation predictable. With the stator polarities indicated, the rotor would align itself midway between the two pairs of stator coils. The direction of rotation can now be predicted and is determined by the polarities of the stator coil sets. Adding more stator coil pairs to a motor of this type improves its rotation and makes the stepping action very accurate. Figure 11-10 shows several types of dc stepping motors.

Figure 11-9 Two sets of stator coils to produce a constant rotor movement.

Figure 11-8 Stator coils with power applied.

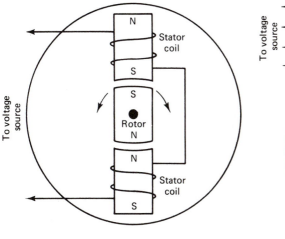

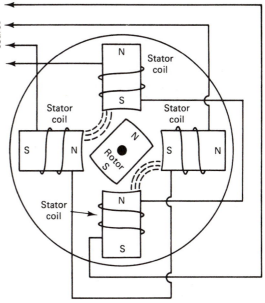

(a)

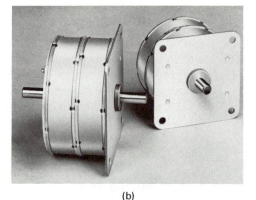
(b)

(c)

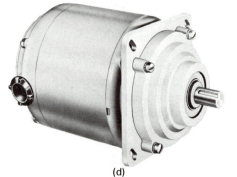
(d)

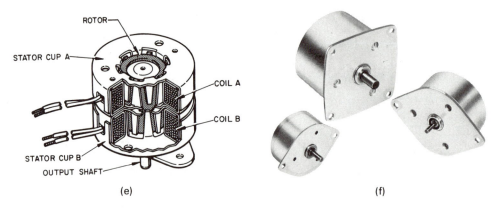

(e)

(f)

Figure 11-10 Types of dc stepping motors. [(a)–(d), Courtesy of Superior Electric Co.; (e), (f), courtesy of Airpax Corp.]

Figure 11-10 shows an electrical diagram and switching sequence of a dc stepping motor. The stator coils of this motor are wound by a special type of construction called *bifilar.* Two separate wires are wound into the coil slots at the same time. The two wires are small in size, allowing more turns than could be achieved with larger wire. Construction of this type simplifies the control circuitry and dc input source requirements to drive the motor.

228

Operation of the stepping motor of Figure 11-11 may be achieved in a four-step switching sequence. Any of the four combinations of switches 1 or 2 will produce a corresponding rotor position location. After the four switch combinations have been achieved, the switching cycle repeats itself. Each switching combination causes the motor to move one-fourth of a step.

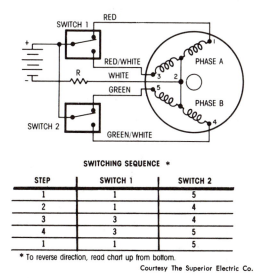

SWITCHING SEQUENCE *

STEP	SWITCH 1	SWITCH 2
1	1	5
2	1	4
3	3	4
4	3	5
1	1	5

* To reverse direction, read chart up from bottom.

Courtesy The Superior Electric Co.

Figure 11-11 Circuit diagram and switching sequence of a bifilar-wound dc stepping motor. (Courtesy of Superior Electric Co.)

A rotor similar to the one shown in Figure 11-6 normally has 50 teeth. Using a 50-tooth rotor in the circuit of Figure 11-11 would permit four steps per tooth, or 200 steps per revolution. The amount of displacement or step angle of this motor is therefore determined by the number of teeth on the rotor and the switching sequence.

A stepping motor that takes 200 steps to produce one revolution will move 360° each 200 steps or 1.8° per step. It is not unusual for stepping motors to use eight switching combinations to achieve one step. In this case, each switching combination could be used to produce 0.9° of displacement. Motors and switching circuits of this type permit a very precise controlled movement. An *eight-step* switching sequence called *half-stepping* is also used. During this type of operation, the motor shaft moves half of its normal step angle for each input pulse applied to the stator. This allows more precise movement control and greater speed capability. Figure 11-12 shows the circuit diagram and switching arrangement of the eight-step switching mode.

Dc stepping motor terminology.   The rotation of a stepping motor shaft is in fixed, repeatable increments (steps). Each time the stator winding polarity is changed by the input logic circuitry, the motor shaft rotates a specific amount. This degree of rotation is called the *step angle* (in degrees). The number of steps required for the shaft to rotate 360° or one complete revolution may be found by dividing the step angle into 360°. Several step angles are available (see Figure 11-13).

Eight-step switching sequence[a]

Step	SW1	SW2	SW3	SW4
1	On	Off	On	Off
2	On	Off	Off	Off
3	On	Off	Off	On
4	Off	Off	Off	On
5	Off	On	Off	On
6	Off	On	Off	Off
7	Off	On	On	Off
8	Off	Off	On	Off
1	On	Off	On	Off

[a] To reverse direction, read chart from the bottom up.

(a)

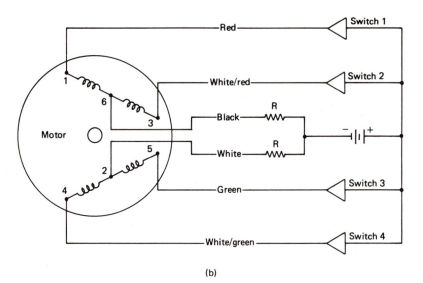

(b)

Figure 11-12 Circuit diagram and switching sequence of an eight-step dc stepping motor. (Courtesy of Superior Electric Co.)

Figure 11-13 Some step angles available.

Step angle (deg)	Steps per 360° rotation
0.72	500
1.8	200
2.0	180
2.5	144
5.0	72

Rather than using an r/min speed rating, as motors with free-running shaft do, the dc stepping motor is rated in *steps per second*. This rating is the number of steps the motor shaft rotates in 1 s. The term *step accuracy* is used to define a position accuracy tolerance. This percentage figure indicates the total possible error of the stepping motor in moving one step. The total error does not accumulate with each additional step. This condition is desirable in precision automated manufacturing systems in industry.

Step response time is another important rating of dc stepping motors. This value indicates the time required for the motor shaft to complete a single step after a dc input signal has been applied to the stator. This time rating is usually in milliseconds and relies on the torque-to-inertia ratio of the motor and the dc input source characteristics.

The *torque-to-inertia ratio* is determined by the following method:

$$\text{torque-to-inertia ratio} = \frac{\text{holding torque (oz-in.)}}{\text{rotor inertia (oz-in. s}^2)}$$

The holding torque of a dc stepping motor is the amount of torque in ounce-inches (oz-in.) required to move the motor shaft from a 0-r/min condition.

DRIVEN LINEAR ACTUATORS

In industry today, there is a need for linear actuators that provide movement changes that can be accurately controlled at different speeds, be made to stop instantly, and to hold a position without further demand for power. Driven linear actuators are designed to achieve these operating conditions. The term "driven," in this case, refers to electric-motor-operated mechanisms that produce linear motion.

An example of driven linear actuators is as follows. When a drive motor is activated, it rotates the threaded drive screw. A mating nut on the drive screw, which is attached to a drive rod, is thereby made to move axially. This action causes the drive rod either to extend or retract, according to the direction of rotation of the drive motor.

When the drive motor of the actuator is set into motion, completion of the stroke or some type of overload will activate an overload cam. This action can be used to shut off the motor or reverse its rotation, according to the design of the actuator. This mechanism can also be triggered by an outside control unit. It will permit operation for certain time periods, or it can be triggered by a series of events. Feedback signals can also be generated to provide accurate positioning control in closed-loop applications. Driven actuators are used wherever linear motion is desired. This includes lever arms, cranks, slides, conveyor belt positioning, and diverter gate control.

The function of a driven linear actuator is fundamentally the same as that of a hydraulic or pneumatic cylinder. Electromagnetic actuators of this type have a number of unique advantages over their fluid counterparts. This includes such things as variable drive velocity, installation at remote locations, unlimited control capabilities, and overload protection.

FREQUENCY CONVERSION SYSTEMS

The power system frequency used in the United States is 60 Hz or 60 cycles per second. However, there are specific applications that require other frequencies in order to operate properly. Mechanical frequency converters are used to change an incoming frequency into some other frequency. Frequency converters are motor-generator sets that are connected together.

For example, a frequency of 60 Hz could be applied to a synchronous motor which rotates at a specific speed. A generator connected to its shaft could have the necessary number of poles to cause it to produce a frequency of 25 Hz. Recall that frequency is determined by the following relationship:

$$\text{frequency (Hz)} = \frac{\text{speed of rotation (r/min)} \times \text{number of poles}}{120}$$

A frequency-conversion system is shown in Figure 11-14. Synchronous units such as the one shown are used wherever precise frequency control is required. It is also possible to design units which are driven by induction motors if some frequency variation can be tolerated.

Figure 11-14 (a) 60-Hz to 25-Hz frequency conversion system; (b) 800-hp variable-frequency drive. [(b), Courtesy of Robicon Corp.]

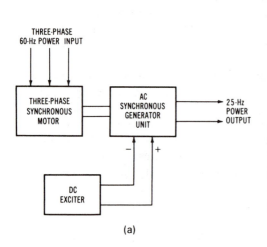

(a)

(b)

Single-phase to three-phase converters.   Two methods are available for converting single-phase ac power to three-phase ac power. The static method involves the use of inductors and capacitors to accomplish phase separation near 120 electrical degrees. A static single-phase to three-phase conversion system is shown in Figure 11-15.

A rotary single-phase to three-phase conversion system is shown in Figure 11-16. This system is a motor-generator unit. A single-phase motor is used to drive a three-phase generator unit. Either the static or rotary system may be used where three-phase power must be obtained from a single-phase system. Machines which use three-phase motors are more economical to operate. In some cases, it is more cost-effective to install a single-phase to three-phase conversion system than to have three-phase power installed in a building.

Figure 11-15   Static single-phase to three-phase conversion system. (Courtesy of Ronk Electrical Industries, Inc.)

Figure 11-16   Rotary single-phase to three-phase conversion system. (Courtesy of Ronk Electrical Industries, Inc.)

GEAR-REDUCTION AND ADJUSTABLE-SPEED MOTORS

Gear-reduction motors, such as the one shown in Figure 11-17, are used where reduced speed of operation is required for a motor. Most gear-reduction motors have either horizontal shafts or shafts at a right angle (90°) to the motor frame. The horsepower delivered at the output shaft of a gear-reduction motor is always less than that produced by the motor. This is due to gear-reduction mechanical losses. Gear-reduction motors should be selected by their output shaft speed and torque ratings rather than horsepower.

An adjustable-speed motor is shown in Figure 11-18. Adjustable speed is obtained by turning a handwheel on top of the motor. The handwheel produces speed changes by changing the diameter of two variable-diameter pulleys which are driven by a heavy-duty belt drive. Each variable-diameter pulley has two disks. One disk of each pulley is movable, allowing adjustable position due to the amount of belt tension.

Figure 11-17 Gear-reduction motor (ac induction type). (Courtesy of Bodine Electric Co.)

Figure 11-18 Adjustable-speed reduction motor. (Courtesy of Litton Industrial Products, Inc., Louis-Allis Division)

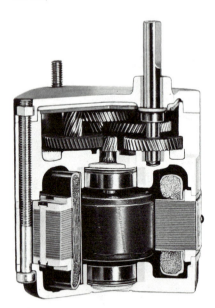

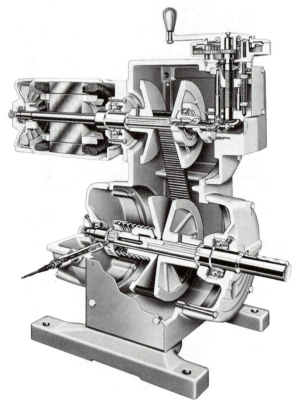

SPECIALIZED MACHINERY FOR ROBOTICS
AND AUTOMATED MANUFACTURING

The development of robotic systems and automated manufacturing in industry has brought about many new applications of specialized electrical machinery. The operation of automated manufacturing systems and industrial robots relies upon various types of electrical machinery. Electrical power systems are depended on to supply energy to manufacturing equipment so that industrial operations and processes may be performed. A robotic system is a unique type of automated manufacturing system. These systems sometimes require several types of energy inputs and specialized machinery for proper operation. Operation of the electrical machinery used with robotic systems is dependent on the proper distribution and control of electrical power. Robotic systems use both alternating-current (ac) and direct-current (dc) electrical energy inputs.

The ac electrical energy requirement for an automated manufacturing system may be either three-phase or single-phase. For larger machines, three-phase ac is ordinarily used. Several machines used for automated manufacturing require direct-current (dc) power supplies. Industries use dc sources for many specialized automated processes. Electroplating and dc variable-speed motor drives are two examples that show the need for dc energy for automated operations. Many robotic systems use dc servomotors and controls. Although most of the electrical power produced in the United States is three-phase alternating current, several methods are available to convert alternating current to direct current for automated manufacturing processes. Direct current is also made available through primary and secondary chemical cells, which are used extensively. The process of converting alternating current to direct current is called *rectification.* Rectification systems are usually the most convenient and inexpensive method to provide dc energy for automated industrial processes.

For automated manufacturing processes today there is a need for devices that produce a type of rotary motion that is somewhat different from that produced by typical electric motors. This type of actuator employs rotary motion to control the angular position of a shaft. Servomechanisms of various types are used in industry to achieve this basic operation for automated manufacturing systems. Through these devices, it becomes possible to transmit rotary motion between locations without direct mechanical linkage. Robotic systems ordinarily use rotary actuators such as dc stepping motors or ac synchronous motors to control their movement. Some industrial robots are shown in Figure 11-19. Figure 11-20 shows some servomotors such as those used with robotic systems.

The power supply of an industrial robot provides energy to operate the actuators of the system. There are three major types of power supplies: (1) electrical, (2) hydraulic, and (3) pneumatic. *Electrical* power supplies require less floor space and provide low-noise operation. *Hydraulic* power supplies may be used to move heavy objects and are faster and more accurate than electrical supplies. *Pneumatic* power supplies are used for light application and have good speed and accuracy. Electrically

(a)

(b)

(c)   **Figure 11-19**   Industrial robots.

(a)   (b)   (c)

Figure 11-20   Servomotors for industrial robots. (Courtesy of Superior Electric Co.)

operated robotic systems are usually driven by dc stepping motors. These systems are not as powerful or as fast in operational speed as hydraulic units; however, they have better accuracy and repeatability and require less floor space. Hydraulic robotic systems have fewer moving parts and are stronger and faster in operation. Pneumatic actuators are ordinarily used for small limited-sequence pick-and-place operations.

REVIEW

11.1. What is a synchro system?

11.2. What is a servomechanism?

11.3. Discuss the operation of a synchro system.

11.4. Discuss the operation of an ac synchronous motor.

11.5. Discuss the operation of a dc stepping motor.

11.6. Describe the construction of a dc stepping motor.

11.7. What is meant by "half-stepping" of a dc stepping motor?

11.8. Define the following terms: **(a)** step angle, **(b)** steps per second, **(c)** step accuracy, **(d)** step response time, and **(e)** torque-to-inertia ratio.

11.9. What is a linear actuator?

11.10. Discuss the operation of a frequency-conversion system.

11.11. What are two types of single-phase to three-phase conversion systems?

11.12. What is the purpose of using a gear-reduction motor?

TWELVE

Electrical Machine Control Systems

The previous chapters have dealt primarily with the types and characteristics of electrical machines. Various types of circuits and equipment are used to control these machines. This chapter provides an overview of several power control systems which are used with electrical machines.

CONTROL SYMBOLS

It is necessary to become familiar with the electrical symbols which are commonly used with machine control systems. Some common machine control symbols are shown in Figure 12-1.

MACHINE CONTROL WITH SWITCHES

An important type of electrical machine control is the switch. Many types of switches are used to control electrical machines. The function of a switch is to turn a circuit on or off; however, many more complex switching functions can be performed.

Toggle switches.   Among the simplest types of switches are toggle switches. The symbols for several kinds of toggle switches are shown in Figure 12-2. Notice the symbols that are used for various types of toggle switches.

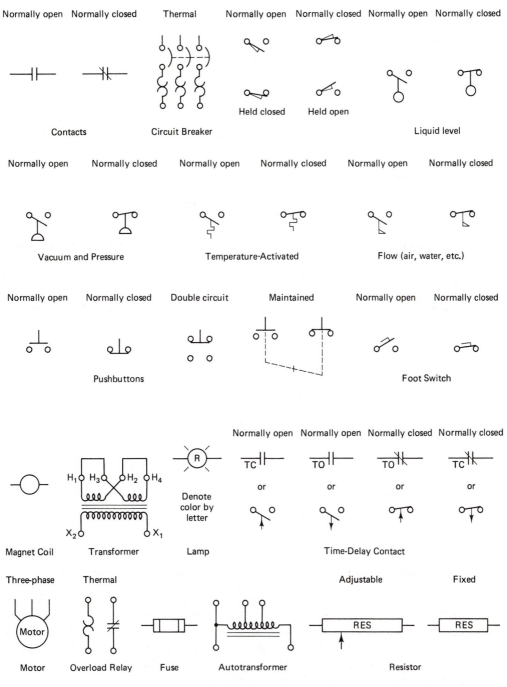

Figure 12-1 Common machine control symbols.

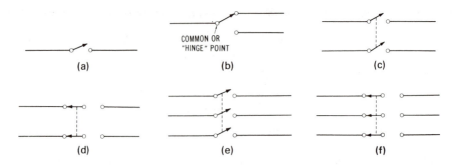

Figure 12-2   Toggle switches: (a) single pole, single throw; (b) three-way, or single pole, double throw; (c) double pole, single throw; (d) double pole, double throw; (e) three pole, single throw; (f) three pole, double throw.

Pushbutton switches.   Pushbutton switches are commonly used for machine control. Many machine control applications use pushbuttons as a means of starting, stopping, or reversing a motor. Pushbuttons are manually operated to close or open a machine control circuit. There are several types of pushbuttons used for the control of machines. Figure 12-3 shows a side view of some pushbutton styles. Pushbuttons are usually mounted in enclosures called motor control stations.

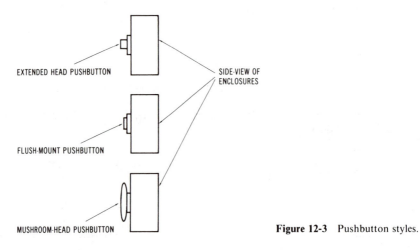

Figure 12-3   Pushbutton styles.

Ordinarily, pushbuttons are either the normally closed (NC) or normally open (NO) types. However, there are a few modifications. A normally closed pushbutton is closed until it is depressed manually and will open a circuit when it is depressed. The normally open pushbutton is open until it is manually depressed and, then, once it is depressed, it will close a circuit. The "start" pushbutton of a motor control station is of NO type, while the "stop" switch is of NC type.

Rotary switches. Another common type of switch is the rotary switch. Many different switching combinations can be wired by using a rotary switch. The shaft of a rotary switch is attached to sets of moving contacts. These moving contacts touch different sets of stationary contacts which are mounted on ceramic segments when the rotary shaft is turned to different positions. The shaft can lock into place in any of several positions. A rotary switch is shown in Figure 12-4. Rotary switches are usually controlled by manually turning the rotary shaft clockwise or counterclockwise. A knob is normally fastened to the end of the rotary shaft to permit easier turning of the shaft.

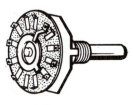

Figure 12-4 Rotary switch.

Limit switches. Some typical limit switches are shown in Figure 12-5. They are made in a variety of sizes. Limit switches are merely on/off switches that use mechanical movement to cause a change in the operation of the electrical control circuit of a machine. The electrical current developed as a result of the mechanical movement is used to *limit* movement of the machine or to make some change in its operational sequence. Limit switches are often used in sequencing, routing, sorting, or counting operations in industry. Often, they are used in conjunction with hydraulic or

Figure 12-5 Limit switches. (Courtesy of Allen-Bradley Co.)

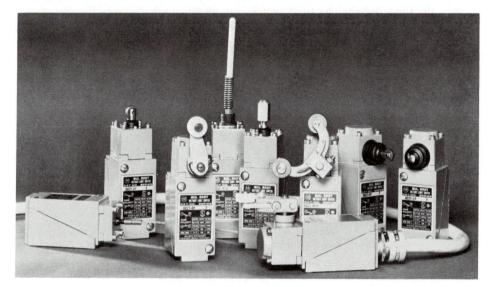

pneumatic controls, electrical relays, or other motor-operated machinery such as drill presses, lathes, or conveyor systems.

In its most basic form, a limit switch converts mechanical motion into an electrical control current. Some limit switches are shown in Figure 12-5. Notice the *cam,* which is an external part that is usually mounted on a machine. The cam applies force to an actuator of the limit switch. The *actuator* is the part of the limit switch that causes the internal NO or NC contacts to change state. The actuator operates due to either linear or rotary motion of the cam, which applies force to the limit switch. The other terms associated with limit switches are pretravel and overtravel. *Pretravel* is the distance that the actuator must move to change the normally open or normally closed state of the limit-switch contacts. *Overtravel* is the distance the actuator moves beyond the point where the contacts change state. Both pretravel and overtravel settings are important in machine setups where limit switches are used.

Temperature switches.   Temperature switches are common types of control devices. The control element of a temperature switch contains a specific amount of liquid. The liquid increases in volume when temperature increases. Thus changes in temperature can be used to change the position of a set of contacts within the temperature-switch enclosure. Temperature switches may be adjusted throughout a range of temperature settings.

Float switches.   Float switches are used when it is necessary to control the level of a liquid. Float switches (see Figure 12-6) usually have an operating lever connected to a rod-and-float assembly. The float assembly is placed into a tank of liquid where the motion of the liquid controls the movement of the operating lever of the float switch. The float switch usually has a set of normally open and normally closed contacts which are controlled by the position of the operating lever. The contacts are often connected to a pump-motor circuit. In operation, the normally open

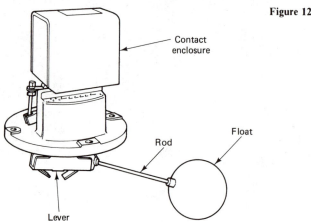

Figure 12-6   Float switch.

contacts could be connected in series with a pump-motor control circuit. When the liquid level is reduced, the float switch would be lowered to a point where the operating lever would be moved far enough that the contacts would be caused to change to a closed state. The closing of the contacts would cause the pump motor to turn on. More liquid would then be pumped into the tank until the liquid level had risen high enough to cause the float switch to turn the pump motor off.

Pressure switches.   Another type of electrical control device is called a pressure switch. A pressure switch has a set of electrical contacts which change states due to a variation in the pressure of air, hydraulic fluid, water, or some other medium. Some pressure switches are diaphragm operated and rely on the intake or expelling of a medium such as air. This action takes place in a diaphragm assembly within the pressure-switch enclosure. Another type of pressure switch uses a piston mechanism to initiate the action of opening or closing the switch contacts. In this type, the movement of a piston is controlled by the pressure of the medium (air, water, etc.).

Foot switches.   A foot switch is a switch that is controlled by a foot pedal. This type of switch is used for applications where a machine operator has to use both hands during the operation of the machine. The foot switch provides an additional control position for the operation of a machine for such times as when the hands cannot be used.

CONTROL EQUIPMENT FOR ELECTRICAL MACHINES

Several types of electromechanical equipment are used for the control of electrical machines. The selection of control equipment affects the efficiency of the system and the performance of the machinery. It is very important to use the proper type of equipment for each machine control application. This section will discuss some types of equipment used for motor control.

Motor Starting Control

A motor starting device is a type of machine control used to accelerate a motor from a "stopped" condition to its normal operating speed. There are many variations in motor starter design, the simplest being a manually operated on/off switch connected in series with one or more power lines. This type of starter is used only for smaller motors which do not draw an excessive amount of current. Several types of motor starter controls are shown in Figure 12-7.

One type of motor starter is the *magnetic starter* which relies on electro-magnetism to open or close the power source circuit of the motor. An example of this type of starter is shown in Figure 12-7(c). Often, motor starters and other control

Figure 12-7 Motor starter controls: (a) across-the-line starter with thermal overload protection shown; (b) starter mounted in an enclosure; (c) starter with three sets of contact lugs; (d) 100-hp ac motor starter; (e) solid-state motor starter. [(a), (b), Courtesy of Eaton Corp., Cutler-Hammer Products; (c), courtesy of Allen-Bradley Co.; (d), courtesy of Robicon Corp.; (e), courtesy of Westinghouse Electric Corp.]

Figure 12-8 Power control center. (Courtesy of Basler Electric Company.)

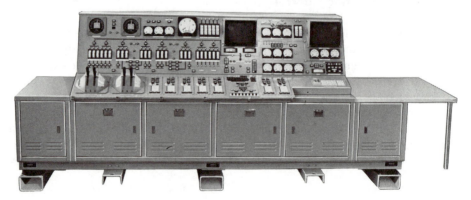

equipment are grouped together for the control of adjacent machines. Such groupings of starters and associated control equipment, such as the one shown in Figure 12-8, are called power control centers. Control centers provide easier access to the power system since they are more compact and the control equipment is not scattered throughout a large area.

Several types of motor starters are used for control of electrical machines. The functions of starters vary in complexity; however, they usually perform one or more of the following functions: (1) on and off control, (2) acceleration, (3) overload protection, or (4) reversing direction of rotation.

Some starters control a motor by being connected directly across the power input lines. Other starters reduce the level of input voltage that is applied to the motor when it is started so as to reduce the value of the starting current. Ordinarily, motor overload protection is contained in the same enclosure as the starter.

Motor contactors.   A typical electric motor contactor circuit is illustrated in Figure 12-9. A contactor is a control element of several types of motor starters that actuates a machine. Pressing the start pushbutton switch in Figure 12-9 completes a

Figure 12-9   Motor contactor circuit.

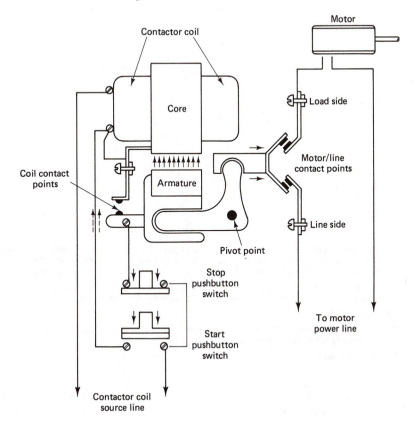

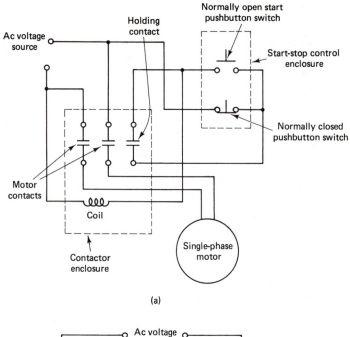

(a)

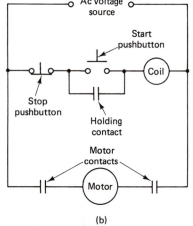

(b)

(1) Press start pushbutton. Circuit is completed when the normally open pushbutton (PB) is momentarily closed. Current flows from voltage source, through normally closed stop PB, through start PB, through the contactor coil, to the other side of the voltage source. This energizes the coil and causes all contacts to close.

(2) Contacts close. Motor will start since ac voltage is applied through the contacts directly to the motor. The holding contact causes the current path through the coil to be complete after the start PB is released.

(3) Motor will run.

(4) Press stop pushbutton. The current path is opened by momentarily opening the normally closed stop PB. The coil deenergizes, causing the contacts to open.

(5) Motor will stop.

(c)

Figure 12-10 Circuit diagram and explanation of a motor contactor circuit: (a) schematic diagram; (b) line diagram; (c) operation.

low-current path to the contactor coil. The contactor coil produces a magnetic field that attracts the armature. Mechanical movement of the armature then completes an electrical path between the power line and the motor through a set of contact points. When this action takes place, the motor starts its operational cycle.

The coil contact points on the left side of the armature are pulled together by the same armature action. As a result, release of the start button at this time will *not* deenergize the contactor coil. A path to the coil source is now completed through the stop pushbutton switch and the coil contact points on the armature. The motor therefore continues to operate as long as electric power is applied. The contact used for this function is called a *holding* contact.

Stopping the operation of a contactor-controlled motor is achieved by pushing the stop button. This action opens the contactor coil power source, which deenergizes the coil and causes the armature to drop out of position. The contactor points in series with the motor also break contact, which deenergizes the power to the motor. This path then becomes incomplete and motor operation stops immediately.

The relay action of a motor contactor is designed to have a latching characteristic that holds it into operation once it is energized. This condition is a necessity in motor control applications. In addition to this, contactors may also be used to actuate a series of operations in a particular sequence. A circuit diagram and brief explanation of the operation of a motor contactor are provided in Figure 12-10.

The contactors used with motor starters are rated according to their current-handling capacity. The National Electrical Manufacturers Association (NEMA) has developed standard sizes for magnetic contactors according to current capacity. Table 12-1 lists the NEMA standard sizes for magnetic contactors. By looking at this chart, you can see that a NEMA size 2 contactor has a 50-A current capacity if it is open (not mounted in a metal enclosure) and a 45-A capacity if it is enclosed. The corresponding maximum horsepower ratings of loads for each of the contactor sizes are also shown in Table 12-1.

TABLE 12-1   SIZES OF MAGNETIC CONTACTORS

NEMA Size	Ampere Rating		Maximum Horsepower Rating of Load					
			Single-Phase		Three-Phase			
	Open	Enclosed	115 V	230 V	115 V	200 V	230 V	460 V
00	10	9	0.33	1	0.75	1.5	1.5	2
0	20	18	1	2	2	3	3	5
1	30	27	2	3	3	7.5	7.5	10
2	50	45	3	7.5	7.5	10	15	25
3	100	90	7.5	15	15	25	30	50
4	150	135	—	—	25	40	50	100
5	300	270	—	—	—	75	100	200
6	600	540	—	—	—	150	200	400
7	900	810	—	—	—	—	300	600
8	1350	1215	—	—	—	—	450	900
9	2500	2250	—	—	—	—	800	1600

Manual starters.   Some electrical machines use *manual starters* to control their operation. This type of starter provides starting, stopping, and overload protection similar to that of a magnetic contactor. However, manual starters are usually mounted near the machine that is being controlled. Remote-control operation is not possible as it is with a magnetic contactor. This is due to the small control current that is required by magnetic contactors. Magnetic contactors also provide low-voltage protection by the drop-out of the contacts when a low-voltage level occurs. Manual starters remain closed until they are manually turned off and are usually limited to small sizes.

A uniform method has also been established for sizing manual starters. Some examples of the sizes of manual starters listed in Table 12-2 are:

1. *Size M-0:* for single-phase 115-V motors up to 1 hp
2. *Size M-1:* for single-phase 115-V motors up to 2 hp
3. *Size M-1P:* for single-phase 115-V motors up to 3 hp

TABLE 12-2   SIZES OF MANUAL STARTERS

Starter Size	Maximum Horsepower Rating of Load				Ampere Rating
	Single-Phase		Three-Phase		
	115 V	230 V	200-230 V	460-575 V	
M-0	1	2	3	5	20
M-1	2	3	7.5	10	30
M-1P	3	5	—	—	30

Classes of motor starters.   Motor starters are divided into five classes which are established by the National Electrical Manufacturers Association (NEMA). These are:

1. *Class A:* alternating-current manual or magnetic starters which operate on 600 V or less
2. *Class B:* direct-current manual or magnetic starters which operate on 600 V or less
3. *Class C:* alternating-current intermediate-voltage starters
4. *Class D:* direct-current intermediate-voltage starters
5. *Class E:* alternating-current magnetic starters which operate on 2200 to 4600 V; Class E1 uses contacts and Class E2 uses fuses

Combination starters.   A popular type of motor starter used in machine control applications is the combination starter. These starters have switching equipment, protective devices, and machine control equipment mounted in a common enclosure. Combination starters are used on systems of 600 V or less. A combination starter is shown in Figure 12-11.

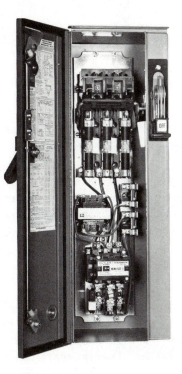

Figure 12-11 Combination starter. (Courtesy of Eaton Corp., Cutler-Hammer Products.)

Selection of motor starters. There are several important criteria to consider when selecting electric motor starters. Among these are:

1. *Type of motor:* ac or dc, induction or wound rotor
2. *Motor ratings:* voltage, current, duty cycle, and service factor
3. *Motor operating conditions:* ambient temperature and type of atmosphere in which it is used
4. *Utility company regulations:* power factor, demand factor, load requirements, and the local electrical codes
5. *Load:* type of mechanical load connected to motor-torque requirements for machine operation

Motor starter enclosures. The purpose of a motor starter enclosure is to protect the operator against accidental contact with high voltages which could cause shock or death. In some cases enclosures are used to protect the control equipment from its operating environment, which may contain water, heavy dust, or combustible materials. The categories of motor controller enclosures were standardized by the National Electrical Manufacturers Association (NEMA). The following is a list of enclosure classifications:

1. *NEMA 1:* general-purpose
2. *NEMA 3 (3R):* weatherproof
3. *NEMA 4:* weathertight

4. *NEMA 4X:* corrosion resistant
5. *NEMA 12:* industrial use
6. *NEMA 13:* oil-tight pushbutton enclosure
7. *NEMA 7:* for atmospheres containing hazardous gas
8. *NEMA 9:* for atmospheres containing explosive dust
9. *NEMA Types 1B1, 1B2, 1B3:* flush types; provide behind-the-panel mounting into machines bases, columns, or plaster walls to conserve space and to provide a more pleasant appearance; NEMA 1B1 mounts into an enclosed machine cavity; NEMA 1B2 includes its own enclosure behind the panel to exclude shavings and chips; NEMA 1B3 is for plaster wall mounting

Relays

Relays are widely used control devices. They have an electromagnet which contains a stationary core. Mounted close to one end of the core is a movable piece of magnetic material called the *armature.* When the coil is activated electrically, it produces a magnetic field around the metal core. The armature is then attracted toward the core. When the coil is deenergized, the armature is returned to its original position by spring action. Figure 12-12 shows a simplified diagram of a relay used to control a motor.

The armature of a relay is generally designed so that electrical contact points respond to its movement. Activation of the relay coil causes the contact points to "make" or "break" according to the design of the relay. A relay is considered to be an electromagnetic switching mechanism. There are many special-purpose relays and switch combinations used for electrical machine control.

Relays use a small amount of current to create an electromagnetic field that is strong enough to attract the armature. When the armature is attracted it either opens or closes the contacts. The contacts, then, either turn on or turn off circuits that are

Figure 12-12   Simplified diagram of the construction of a relay that is used to control a motor.

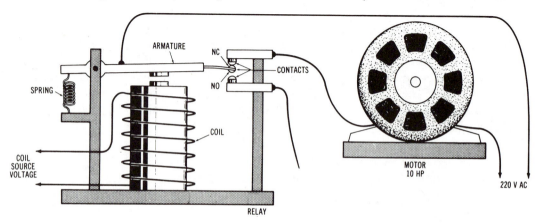

using large amounts of current. The minimum current that flows through the relay coil in order to create a magnetic field strong enough to attract the armature is known as the *pickup current*. The current through the relay coil that allows the magnetic field to become weak enough to release the armature is known as the *dropout current*.

There are two types of contacts used in conjunction with most relays: normally open and normally closed. The normally open contacts remain open when the relay coil is deenergized and closes when the relay is energized. The normally closed contacts remain closed when the relay is deenergized and open when the coil is energized.

Solenoids

A solenoid, shown in Figure 12-13, is an electromagnetic coil with a movable core that is constructed of a magnetic material. The core, or plunger, is sometimes attached to an external spring. This spring causes the plunger to remain in a fixed position until moved by the electromagnetic field that is created by current through the coil. This external spring also causes the core or plunger to return to its original position when the coil is deenergized.

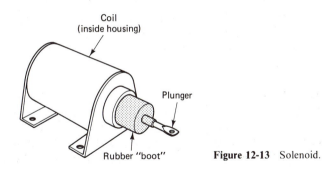

Coil
(inside housing)

Plunger

Rubber "boot"   **Figure 12-13** Solenoid.

Solenoids are used for a variety of control applications. Many gas and fuel-oil furnaces use solenoid valves to turn the fuel supply on or off automatically upon demand. Most dishwashers use one or more solenoids to control the flow of water.

Special-Purpose Relays

There are many special-purpose relays used for electrical machine control. *General-purpose relays* are the type used for low-power applications. They are relatively inexpensive and small in size. Many small general-purpose relays are mounted in octal-base (8-pin) plug-in sockets. *Latching relays* are another type of relay which have a latching mechanism which holds the contacts in position after the power has been removed from the coil. *Solid-state relays* are electronically operated and used where improved reliability or a rapid rate of operation is necessary. Electromagnetic relays will wear out after prolonged use and have to be replaced periodically. Solid-state relays have a longer life expectancy and are not sensitive to shock, vibration, dust, moisture, or corrosion. *Timing relays* are used to turn a load device on or off after a specific period of time. One popular type is a pneumatic timing relay. The operation of a pneumatic timing relay is dependent on the movement of air within a chamber. Air movement is controlled by an adjustable orifice that regulates the rate of air movement through the chamber. The airflow rate determines the rate of movement of a diaphragm or piston assembly. This assembly is connected to the contacts of the relay. Therefore, the orifice adjustment controls the airflow rate, which determines the time from the activation of the relay until a load connected to it is turned on or off. There are other types of timing relays, such as solid-state, thermal, oil-filled, dashpot, and motor-driven timers. Timing relays are useful for sequencing operations where a time delay is required between operations. A typical application is as follows: (1) a start pushbutton is pressed; (2) a timing relay is activated; (3) after a 10-s time delay, a motor is turned on.

MACHINE CONTROL SYSTEMS

Electrical control systems are used with many types of machines or loads. The most common loads are electric motor; however, many of the basic control systems discussed in this chapter are also used to control lighting and heating loads. Generally, the controls for lighting and heating loads are less complex than electric motor control systems.

Start-stop control. Figure 12-14 is a start–stop pushbutton control circuit with overload protection (OL). Notice that the start pushbutton is normally open (NO) and the stop pushbutton is normally closed (NC). Single-phase lines L1 and L2 are connected across the control circuit. When the start pushbutton is pushed, momentary contact is made between points 1 and 2. This causes the NO contact (M) to close. A complete circuit between L1 and L2 causes the electromagnetic coil M to be energized. When the normally closed stop pushbutton is pressed, the circuit between L1 and L2 opens. This causes contact M to open and turn the circuit off.

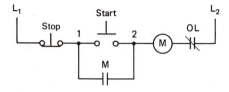

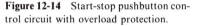

Figure 12-14 Start-stop pushbutton control circuit with overload protection.

This control circuit and the others which are discussed are shown as control diagrams. They illustrate only the control portion of the system. The control diagram shown in Figure 12-14 could be used to control either a single-phase, three-phase, or direct-current machine. The power lines of the system must connect, through a set of contacts, to the line terminals of the machine being controlled. Since this operation is evident for all machines, control diagrams omit this portion of the control system.

Multiple-location start-stop control.   The circuit of Figure 12-15 is a multiple-location start–stop control. In this circuit, the start–stop control of a machine can be accomplished from three separate locations. Notice that the start pushbuttons are connected in parallel and the stop pushbuttons are connected in series. The control of one load from as many locations as is desired can be accomplished with this type of control circuit.

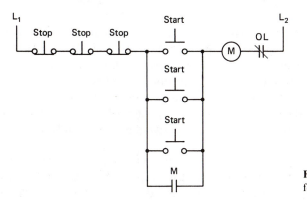

Figure 12-15 Start-stop control circuit for three locations.

Start-stop jog control.   Figure 12-16 has a "jog–run" switch added in series with the normally open contact (M). In the "run" position, the circuit operates just like the start–stop control circuit of Figure 12-14. The "jog" position is used so that a complete circuit between L1 and L2 is achieved and sustained only while the start pushbutton is pressed. With the selector switch in the jog position, a motor can be rotated a small amount at a time for positioning purposes. *Jogging* or *inching* is defined as the momentary operation of a motor to provide small movements of its shaft.

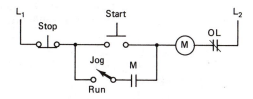

Figure 12-16 Start-stop pushbutton control circuit with a jog–run selector switch.

Figure 12-17 shows another method of motor jogging control. This circuit has a separate pushbutton for jogging which relies on a normally open contact (CR) to operate. Two control relays are used with this circuit. The jog pushbutton energizes coil (M) only, while the start pushbutton energizes coils (CR) and (M) to cause the controlled machine to continue operation.

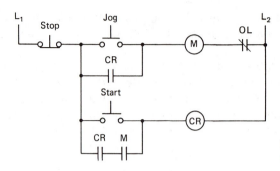

Figure 12-17 Pushbutton control circuit for start–stop jogging.

Forward-reverse-stop control.   The circuit of Figure 12-18 has forward and reverse pushbuttons arranged in sets. When the forward pushbutton is pressed, the top pushbutton will momentarily close and the lower pushbutton will momentarily open. Current then flows from L1 to L2 through coil (F). When coil (F) is energized, normally open contact F will close and the normally closed contact F will open. The "forward" coil will then remain energized. The reverse pushbuttons cause a similar action of the reversing circuit. Two coils, one for forward operation and one for reverse operation, are required.

The circuit of Figure 12-19 is similar in function to that of the circuit in Figure 12-18 except that the pushbutton arrangement is simpler. When the normally open forward pushbutton is pressed, the current through coil (F) is interrupted. When the

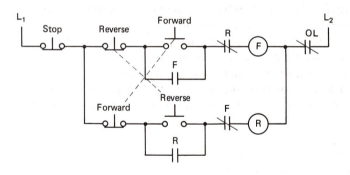

Figure 12-18 Forward–reverse–stop pushbutton control circuit.

Figure 12-19 Forward–reverse–stop pushbutton control circuit.

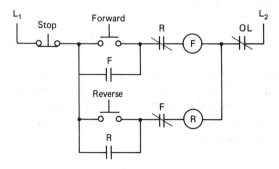

normally open reverse pushbutton is pressed, current flows through coil (R). When the reverse coil is energized, normally open contact F closes and normally closed contact F opens. This action causes a motor to operate in the reverse direction, until the stop pushbutton is pressed again.

There are many other pushbutton combinations which can be used with motor contactors and control relays to accomplish motor control or control of other types of loads. The circuits discussed in this section represent some basic machine reverse–stop and jogging.

MOTOR STARTING SYSTEMS

Motor starting, particularly for large motors, plays an important role in the efficient operation of electrical machinery. Several different systems are used to start electric motors. The motor starting equipment that is used is placed between the electrical power source and the motor. Electric motors draw a larger current from the power source during starting than during normal operation. Motor starting equipment is often used to reduce starting currents to a level that can be handled by the electrical power system.

Full-voltage starting.   One method of starting electric motors is called full-voltage starting. This method is the least expensive and the simplest to install. Since full power supply voltage is applied to the motor initially, maximum starting torque and minimum acceleration time result. However, the electrical power system must be able to handle the starting current drawn by the motor.

Full-voltage starting is illustrated by the diagram of Figure 12-20. In this motor control circuit, a start–stop pushbutton station is used to control a three-phase motor. When the normally open start pushbutton is pressed, current flows through the relay coil (M), causing the normally open contacts to close. The line contacts allow full voltage to be applied to the motor when they are closed. When the start pushbutton is released, the relay coil remains energized due to the holding contact. This contact provides a current path from L1 through the normally closed stop pushbutton, through the holding contact, through the coil (M), through a thermal overload relay, and back to L2. When the stop pushbutton is pressed, this circuit is opened causing the coil to be deenergized.

Primary-resistance starting.   Another motor starting method is called primary resistance starting. This method uses large resistors in series with the power lines to reduce the motor starting current. Often, the resistance connected into the power lines is reduced in steps until full voltage is applied to the motor. Thus starting current is reduced according to the value of the series resistance in the power lines since starting torque is reduced according to the magnitude of current flow.

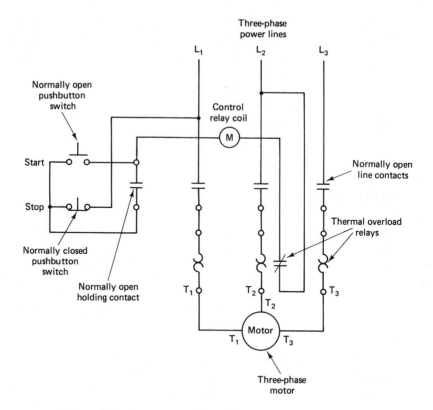

Figure 12-20 Full-voltage starting circuit for a three-phase motor.

Figure 12-21 shows the primary-resistance starting method used to control a three-phase motor. When the start pushbutton is pressed, coils (S) and (TR) are energized. Initially, the start contacts (S) will close, applying voltage through the primary resistors to the motor. These resistors reduce the value of starting current. Once the time-delay period of the timing relay (TR) has elapsed, contact TR will close. The run contacts (R) will then close and apply full voltage to the motor.

Primary-reactor starting.   Another starting method, similar to primary-resistance starting, is the primary-reactor starting method. Reactors (coils) are used in place of resistors since they consume smaller amounts of power from the ac source. Usually, this method is more appropriate for large motors that are rated at over 600 V.

Autotransformer starting.   Autotransformer starting is another method used to start electric motors. This method employs one or more autotransformers to control the voltage that is applied to a motor. The autotransformers used are ordinarily tapped to provide a range of starting-current control. When the motor has accelerated to near its normal operating speed, the autotransformer windings are

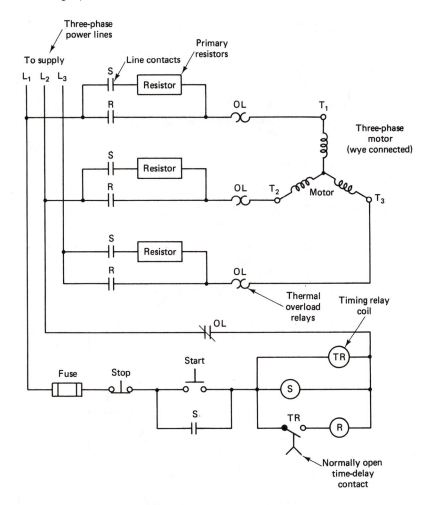

Figure 12-21   Primary-resistance starter circuit.

removed from the circuit. A major disadvantage of this method is the expense of the autotransformers.

An autotransformer starting circuit is shown in Figure 12-22. This is an expensive type of control that uses three autotransformers and four relays. When the start pushbutton is pressed, current flows through coils (1S), (2S), and (TR). The 1S and 2S contacts will then close. Voltage is applied through the autotransformer windings to the three-phase motor. One normally closed and one normally open contact are controlled by timing relay TR. When the specified time period has elapsed, the normally closed TR contact will open and the normally open TR contact will close. Coil (R) then energizes, causing the normally open R contacts to close and apply full voltage to the motor. Normally closed R contacts are connected in series with coils

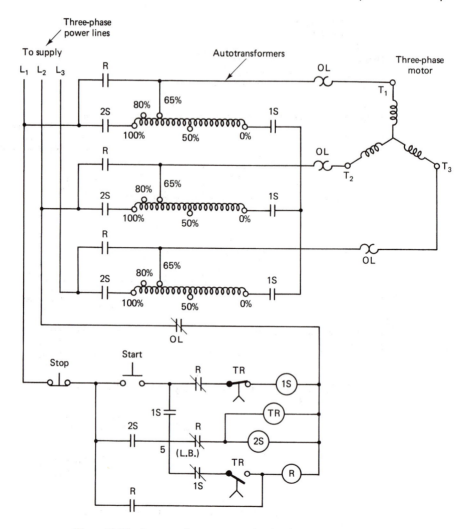

Figure 12-22 Autotransformer starter circuit with a three-phase motor.

(1S), (2S), and (TR) to open their circuits when coil (R) is energized. When the stop pushbutton is pressed, the current to coil (R) is interrupted, thus opening the power-line connections to the motor.

Notice that the 65% taps of the autotransformer are used in Figure 12-22. There are also taps for 50%, 80%, and 100%, to provide more flexibility in reducing the motor-starting current.

Wye-delta starting. It is possible to start three-phase motors more economically by using the wye-delta starting method. Since in a wye configuration, line current is equal to the phase current divided by 1.73 (or $\sqrt{3}$), it is possible to reduce the starting current by using a wye connection rather than a delta connection. This method, shown in Figure 12-23, employs a switching arrangement which places the motor stator windings in a wye configuration during starting and a delta arrangement for running. In this way, starting current is reduced. Although starting torque is reduced, running torque is still high since full voltage appears across each winding when the motor is connected in a delta configuration.

When the start pushbutton in Figure 12-23 is pressed, coil (S) is energized. The normally open S contacts then close. This action connects the motor windings in a wye configuration and also activates timing relay (TR) and coil (1M). The normally open 1M contacts then close to apply voltage to the wye-connected motor windings. After the time-delay period has elapsed, the TR contacts change state. Coil (S) deenergizes and coil (2M) energizes. The S contacts which hold the motor windings in a wye arrangement then open. The 2M contacts then close and cause the motor windings to be connected in a delta configuration. The motor will then continue to run with the stator windings connected in a delta arrangement.

Figure 12-23   Three-phase wye-delta starting circuit.

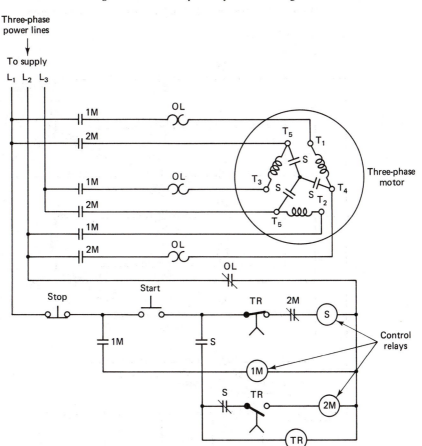

Part-winding starting. Figure 12-24 shows the part-winding-starting method, which is more simple and less expensive than most other starting methods. However, motors must be specifically designed to operate in this manner. During starting, the power-line voltage is applied across only part of the parallel-connected motor windings, thus reducing starting current. Once the motor has started, the line voltage is placed across all of the motor windings. This method is undesirable for many heavy-load applications due to the reduction of starting torque.

In Figure 12-24, when the start pushbutton is pressed, current flows through coil (M1) of the time-delay relay. This causes the normally open contacts of M1 to close and three-phase voltage is applied to windings T1, T2, and T3. After the time-delay period has elapsed, the normally open contact located below coil (M1) closes. This action energizes coil (M2) and causes its normally open contacts to close. The M2 contacts then connect the T7, T8, and T9 windings in parallel with the T1, T2, and T3 windings. When the stop pushbutton is pressed, coils (M1) and (M2) are deenergized.

Figure 12-24   Part-winding circuit for three-phase motor starting.

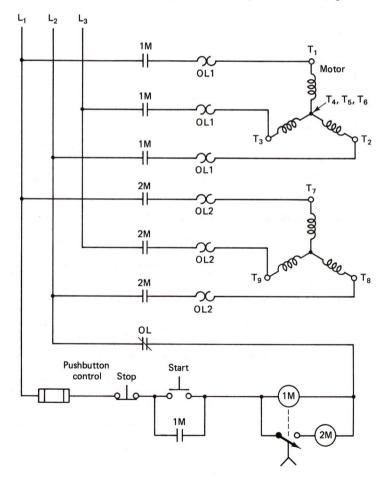

Dc starting systems.   Since dc motors have no counterelectromotive force (CEMF) when they are not rotating, they have tremendously high starting currents. Therefore, they must use some type of control system to reduce the initial starting current, such as a series resistance. Resistance can be manually or automatically reduced until full voltage is applied. The four types of control systems commonly used with dc motors are (1) current limit, (2) definite time, (3) CEMF, and (4) variable voltage. The current-limit method allows the starting current to be reduced to a specified level and then advanced to the next resistance step. The definite-time method causes the motor to increase speed in timed intervals with no regard to the amount of armature current or to the speed of the motor. The CEMF method samples the amount of CEMF generated by the armature of the motor to reduce the series resistance accordingly. This method can be used effectively since CEMF is proportional to both the speed and the armature current of a dc motor. The variable-voltage method employs a variable dc power source to apply a reduced voltage to the motor initially and then gradually increase the voltage. No series resistances are needed when the variable-voltage method is used.

FORWARD AND REVERSE CONTROL

Most types of electrical motors can be made to rotate in either direction by some simple modifications of their winding connections. Ordinarily, motors require two magnetic motor contactors, such as those shown in Figure 12-25, to accomplish forward and reverse operation. These contactors are used in conjunction with a set of three-pushbutton switches: forward, reverse, and stop. When the forward pushbutton switch is depressed, the forward contactor is energized. It is deactivated when the stop pushbutton switch is depressed. A similar procedure takes place during reverse operation.

Figure 12-25 Forward and reverse motor contactors. (Courtesy of Eaton Corp., Cutler-Hammer Products.)

Dc motor reversing. Direct-current motors have their direction of rotation reversed by changing either the armature connections *or* the field connections to the power source. In Figure 12-26, a dc shunt motor control circuit is shown. When the forward pushbutton is pressed, coil (F) is energized, causing the F contacts to close. The armature circuit is then completed from L1 through the lower F contact, *up* through the armature, through the upper F contact, and back to L2. Pressing the stop pushbutton deenergizes coil (F).

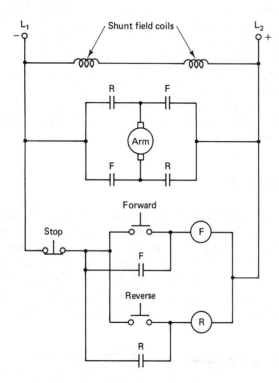

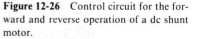

Figure 12-26   Control circuit for the forward and reverse operation of a dc shunt motor.

The direction of rotation of the motor is reversed when the reverse pushbutton is pressed. This is due to the change of the current direction through the armature. Pressing the reverse pushbutton energizes coil (R) and closes the R contacts. The armature current path is then from L1 through the upper R contact, *down* through the armature, through the lower R contact, and back to L2. Pressing the stop button deenergizes coil (R).

Single-phase induction-motor reversing. Single-phase ac induction motors that have start and run windings have their direction of rotation reversed by using the circuit in Figure 12-26. The diagram is modified by replacing the shunt field coils with the run windings and the armature with the start windings. Single-phase induction motors are reversed by changing the connections of either the start windings *or* the run windings but not both at the same time.

Three-phase induction-motor reversing. Three-phase motors have their direction of rotation reversed by simply changing the connections of any two power lines. This changes the phase sequence applied to the motor. A control circuit for three-phase induction-motor reversing is shown in Figure 12-27.

When the forward pushbutton is pressed, the forward coil will energize and close the F contacts. The three-phase voltage is applied from L1 to T1, L2 to T2, and L3 to T3 to cause the motor to operate. The stop pushbutton deenergizes the forward coil. When the reverse pushbutton is pressed, the reverse coil is energized and the R contacts will close. The voltage is then applied from L1 to T3, L2 to T2, and L3 to T1. This action reverses the L1 and L3 connections to the motor and causes the motor to rotate in the reverse direction.

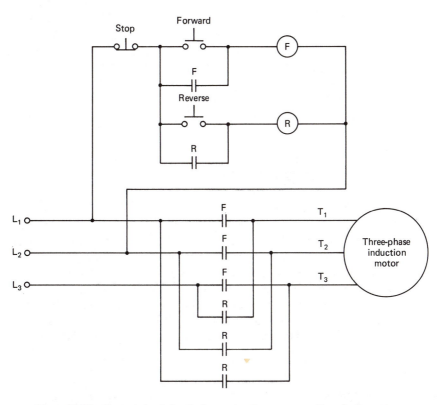

Figure 12-27 Control circuit for the forward and reverse operation of a three-phase induction motor.

DYNAMIC BRAKING

When a motor is turned off, its shaft continues to rotate for a short period of time. This continued rotation is undesirable for many applications. Dynamic braking is a method used to bring a motor to a quick stop whenever power is turned off. Motors with wound armatures utilize a resistance connected across the armature as a dynamic braking method. When power is turned off, the resistance is connection across the armature. This causes the armature to act as a loaded generator, making the motor slow down immediately. This dynamic braking method is shown in Figure 12-28.

Alternating-current induction motors can be slowed down rapidly by placing a dc voltage across the winding of the motor. This dc voltage sets up a constant magnetic field which causes the rotor to slow down rapidly. A circuit for the dynamic braking of a single-phase ac induction motor is shown in Figure 12-29.

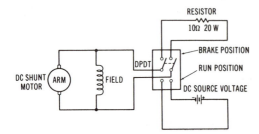

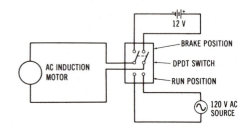

Figure 12-28 Dynamic braking circuit for a dc shunt motor.

Figure 12-29 Dynamic braking circuit for a single-phase ac induction motor.

UNIVERSAL MOTOR SPEED CONTROL

An important advantage of a universal motor is its ease of speed control. The universal motor has a brush/commutator assembly, with the armature circuit connected in series with the field windings. By varying the voltage applied to universal motors, their speed may be varied from zero to maximum.

The circuit used for this purpose, shown in Figure 12-30, uses a gate-controlled triac. The triac is a semiconductor device whose conduction may be varied by a trigger voltage applied to its gate. A silicon-controlled rectifier (SCR) could also be used as a speed-control device for a univeral motor. Speed-control circuits like this one are used for many applications, such as electric drills, sewing machines, electric mixers, and industrial applications.

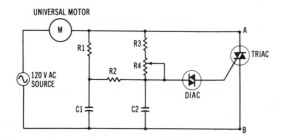

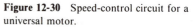

Figure 12-30 Speed-control circuit for a universal motor.

MOTOR OVERLOAD PROTECTIVE DEVICES

Basic functions that motors are expected to perform are starting, stopping, reversing, and speed variation. These functions may be manually or automatically controlled. Various types of protective devices are used for efficient distribution of power to electric motors.

Overload protection is the most important motor protective function. Such protection should serve the motor, its branch circuit, and associated control equipment. The major cause of motor overload is an excessive mechanical load on the motor, which causes it to draw more current from the source.

Thermal-overload relays, shown in Figure 12-31, are often used as protective devices. Thermal relays may be reset either manually or automatically. One type of thermal overload relay uses a bimetallic heater element. The element bends as it is heated by the current through it. When the current reaches the rating of the element, the relay opens the branch circuit. Another type of element is the melting-alloy type. This device has contacts held closed by a ratchet wheel. At the rated current capacity of the device, the fusible alloy melts, causing the ratchet wheel to turn. A spring then causes the device to open the circuit.

Figure 12-31 Melting-alloy thermal overload. (Courtesy of Furnas Electric Co.)

REVIEW

12.1. List some types of toggle switches.

12.2. What are some styles of pushbutton switches?

12.3. What is a limit switch?

12.4. Discuss the operation of a temperature switch.

12.5. Discuss the operation of a float switch.

12.6. Discuss the operation of a pressure switch.

12.7. What is a magnetic motor contactor?

12.8. What is a holding contact?

12.9. What is a manual starter?

12.10. What is a combination starter?

12.11. What are the categories of motor starter enclosures?

12.12. What is a relay?

12.13. What is meant by pickup current? Dropout current?

12.14. What is a solenoid?

12.15. What is a jog control?

12.16. How is reverse control of a three-phase motor accomplished?

12.17. What is meant by the following control terms: **(a)** full-voltage starting, **(b)** primary resistance starting, **(c)** primary reactor starting, **(d)** autotransformer starting, **(e)** wye-delta starting, and **(f)** part-winding starting?

12.18. Why must dc motors use resistance-starting circuits?

12.19. How is reversal of direction of rotation of a single-phase ac induction motor accomplished?

12.20. What is meant by dynamic braking?

12.21. How is universal motor speed control accomplished?

12.22. What is the function of a motor overload protective device?

THIRTEEN

Electrical Machinery Measurement and Testing Systems

The operation and maintenance of electrical machines relies on several types of measurement and testing sytems. Some of the types of equipment commonly used for electrical machine measurements and tests are discussed in this chapter.

BASIC TEST EQUIPMENT

One of the most valuable and often used types of equipment is the volt-ohm-milliammeter (VOM). Two types of VOMs are shown in Figure 13-1. These instruments are used to measure ac voltage, dc voltage, dc current, and resistance. Basic tests on machines may be accomplished using a VOM. It is a very versatile meter which may be used easily for maintenance tests.

MEASUREMENT OF POWER QUANTITIES

Several types of machinery measurement applications involve equipment used to measure electrical power and related quantities. The day-to-day operation of machines relies on the use of electrical power. Therefore, to use electrical machines more effectively, their power conversion should be monitored.

Measuring electrical power and energy. Wattmeters, which monitor both voltage and current in a circuit, are ordinarily used to measure the power converted by electrical machines. For instance, the amount of electrical energy converted into

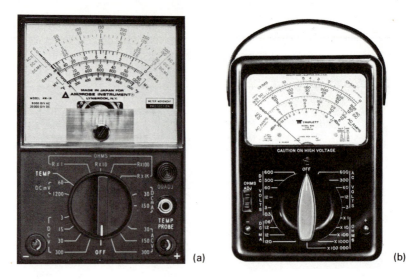

Figure 13–1 Volt-ohm-milliammeters. [(a), Courtesy of Amprobe Instrument, Division of Core Industries Inc; (b), courtesy of Triplett Corp.]

mechanical energy by a motor can be measured by properly connecting a wattmeter to the motor circuit. The meters shown in Figure 13-2 are used to measure electrical power. Figure 13-2(a) shows a meter that is clamped onto an ac power line and Figure 13-2(b) shows a more sophisticated power analyzer.

The amount of electrical energy used over a period of time can be measured by using a watthour meter. A watthour meter, illustrated in Figure 13-3(a), relies on the operation of a small motor mounted inside its enclosure. The speed of the motor is proportional to the power applied to it. The internal construction is shown in Figure 13-3(b). The rotor is an aluminum disk that is connected to a register, which usually

Figure 13-2 Meters used to measure electrical power: (a) clamp-on digital wattmeter; (b) power analyzer. [(a), Courtesy of TIF Instruments, Inc.; (b), courtesy of Magtrol, Inc.]

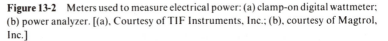

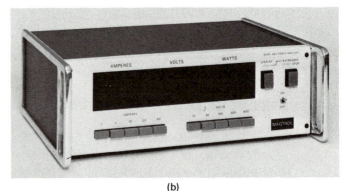

(b)

(a)

(a)

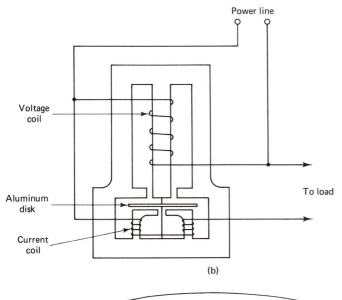

(b)

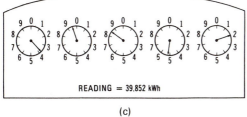

READING = 39,852 kWh

(c)

Figure 13-3 Watthour meter: (a) pictorial; (b) internal construction; (c) dial-type register. [(a) Courtesy of General Electric Co.]

indicates the number of kilowatthours of electrical energy used. Figure 13-3(c) shows the dial type of register that is frequently used on watthour meters. Other types have a direct numerical readout of kilowatthours used.

The watthour meter is connected between the voltage feeder lines and the branch circuits of a power delivery system. In this way, all electrical energy used must pass through the kWH meter.

The electrical operation of a watthour meter relies on a potential coil connected across the power lines to monitor voltage, while a current coil is placed in series with the line to measure current. The voltage and current of the system affect the movement of an aluminum-disk rotor. The operation of the watthour meter is similar to that of an ac induction motor. The stator is an electromagnet that has two sets of windings: the voltage windings and the current windings. The field developed in the voltage windings causes eddy currents to be induced into the aluminum disk. The torque produced is proportional to the voltage and the in-phase current of the system. Therefore, the watthour meter will monitor the *true power* converted.

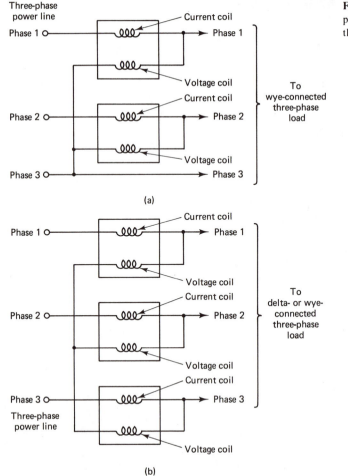

Figure 13-4 Measuring three-phase power: (a) two-wattmeter method; (b) three-wattmeter method.

Measuring three-phase electrical power. For certain applications it is usually necessary to monitor the three-phase power converted by electrical machines. It is possible to use a combination of single-phase wattmeters to measure total three-phase power, as shown in Figure 13-4. The methods shown are ordinarily not very practical since the sum of the meter readings would have to be found in order to calculate the total power of a three-phase system.

Three-phase power meters (see Figure 13-5) are designed to monitor the true power of a three-phase system by using one meter. The reading developed on a power meter is dependent on the voltage and current of the three phases acting on the system. Three-phase watthour meters are also available that use three aluminum disks on the same shaft to monitor the three-phase power used by a system. The total three-phase power is a combination of the power values developed by each phase. Power analyzers such as the clamp-on type shown in Figure 13-5(a) simplify the measurement of three-phase power.

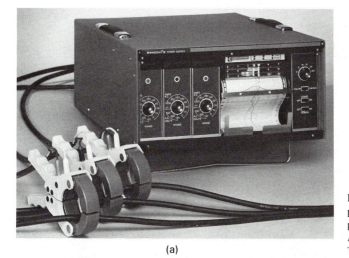

(a)

Figure 13-5 (a) Chart-recording three-phase power analyzer; (b) three-phase power monitors. [(a), Courtesy of Esterline Angus Instrument Corp.; (b), courtesy of Time Mark Corp.]

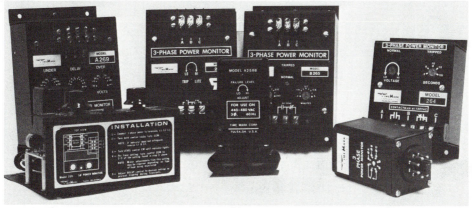

(b)

Measuring power factor.   Power factor is the ratio of the true power of a system to the apparent power (volts × amperes). To determine power factor, a wattmeter, a voltmeter, and an ammeter may be used. The relationship $PF = W/VA$ of each measurement could then be calculated. However, it would be more convenient to use a power factor meter in situations where power factor must be monitored.

The principle of a power-factor meter is illustrated in Figure 13-6. The power factor meter is similar to a wattmeter, except that it has two armature coils that will rotate due to their electromagnetic field strengths. The armature coils are mounted on the same shaft so that their alignment is about 90° apart. One coil is connected across the ac line in series with a resistance, while the other coil is connected across the line in series with an inductance. The resistive path through the coil reacts to produce a flux proportional to the in-phase component of the power. The inductive path reacts in proportion to the out-of-phase component of the power.

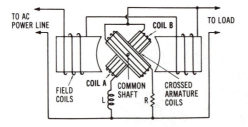

Figure 13-6 Schematic diagram of a power-factor meter.

If a unity (1.0) power factor load is connected to the meter, the current in the resistive path through coil A should develop full torque. Since there is no out-of-phase component, no torque would be developed through the inductive path. The meter movement would now indicate full-scale, or unity power factor. As power factor decreases below 1.0, the torque developed by the inductive path through coil B would become greater. This torque would be in opposition to the torque developed by the resistive path. Therefore, a power factor of less than 1.0 would be indicated. The scale

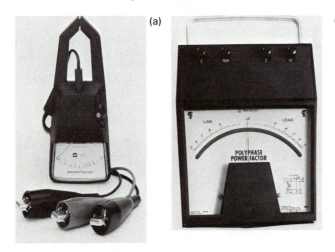

Figure 13-7 (a) Power-factor meter; (b) three-phase power-factor meter. [(a), Courtesy of TIF Instruments; (b), courtesy of Westinghouse Electric Corp.)

must be calibrated to measure power factor ranges from zero to unity. Single-phase and three-phase power factor meters are shown in Figure 13-7.

Various other types of instruments are commercially available for measuring three-phase power and associated quantities. Power-monitoring instruments, which may be used for analysis of true three-phase power, reactive power, current, and voltage, are sometimes referred to as power analyzers [Figure 13-5(a)]. Such equipment usually has several instrument circuits housed in a single enclosure. In this way, the related three-phase quantities can be observed easily to better understand how the power system or machinery is functioning.

MEASUREMENT OF FREQUENCY

Another important measurement is frequency. The frequency of the power source must remain stable or the operation of many types of machines will be affected. Frequency refers to the number of cycles of voltage or current that occur in a given period of time. The international unit of measurement for frequency is the hertz (Hz), which means cycles per second. The standard power frequency in the United States is 60 Hz.

Frequency can be measured with several different types of indicators. An electronic counter is one type of frequency indicator. Vibrating-reed frequency indicators are also common for measuring power frequencies. An oscilloscope, such as the one shown in Figure 13-8, can also be used to measure frequency. Graphic recording instruments can be used to provide a visual display of frequency for different time periods.

Figure 13-8 Oscilloscope in use. (Courtesy of TRW Resistive Products Division.)

SYNCHROSCOPE

The major application of a synchroscope is for generator synchronization power plants. Most power plants have more than one generator (alternator). In order to connect two or more alternators onto the same ac line, the following conditions must be met: (1) their voltage outputs must be equal, (2) their frequencies must be equal, (3) their voltages must be in phase, and (4) the phase sequence of the voltages must be the same.

Voltage output levels can be checked easily with a voltmeter of the step-down transformer type used to monitor high voltages. Frequency is adjusted by varying the speed of the alternator and is also easy to monitor. The phase sequence is established on each alternator when it is installed and connected to the power system. The additional factor, which must also be monitored before paralleling alternators, is to assure that the voltages of the alternator to be paralleled with the other alternators is in phase. This is done with a synchroscope (see Figure 13-9).

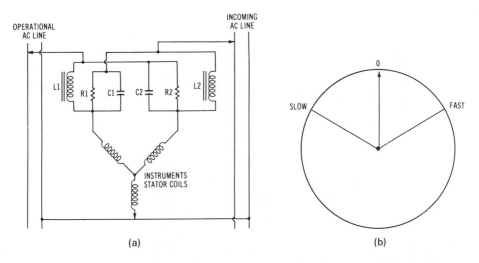

Figure 13-9   Synchroscope: (a) schematic diagram; (b) indicator scale.

The synchroscope is used to measure the relationship between the phase of the system and the alternator to be put "on line." It also indicates whether the alternator is running faster or slower than the system to which it is being connected. The basic design of the indicator utilizes a phase-comparative network of two *RLC* circuits connected between the operating system and the alternator to be paralleled. The indicator scale shows whether the incoming alternator is running too slow or too fast [see Figure 13-9(b)]. When an in-phase relationship exists, along with the other three factors, the alternator can be connected with the system that is already operational. The addition of another alternator will allow a higher power capacity to be produced by the power system.

MEGOHMMETER

A megohmmeter, such as the one shown in Figure 13-10, is used to measure high resistances which extend beyond the range of a typical ohmmeter. These indicators are used primarily for checking the quality of insulation on electrical machines. The quality of insulation of equipment varies with age, moisture content, and applied voltage. The megohmmeter is similar to a typical ohmmeter except that it uses a hand-crank, permanent-magnet dc generator as a voltage source rather than a battery. The dc generator is cranked by the operator while making an insulation test. Figure 13-10 shows a diagram of a megohmmeter circuit. This circuit is essentially the same as any series ohmmeter with the exception of the dc generator used as a voltage source.

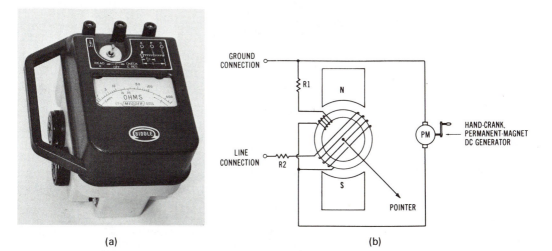

(a)   (b)

Figure 13-10   Megohmmeter: (a) pictorial; (b) schematic diagram. [(a), Courtesy of Biddle Instruments.]

Periodic insulation tests should be made on rotating electrical machines and transformers. As insulation breaks down with age, the equipment becomes malfunctional. A good method is to develop a periodic schedule for checking and recording insulation resistance so that it can be predicted when a piece of equipment needs to be replaced or repaired. A resistance versus time graph can be made and the trend on the graph noted. A downward trend (decrease in insulation resistance) over a time period indicates that a problem in insulation exists.

CLAMP-TYPE METER

Clamp-type meters, such as the ones shown in Figure 13-11, are popular for measuring current in power lines. These indicators can be used for periodic checks of current by clamping them around power lines. They are easy-to-use and convenient maintenance

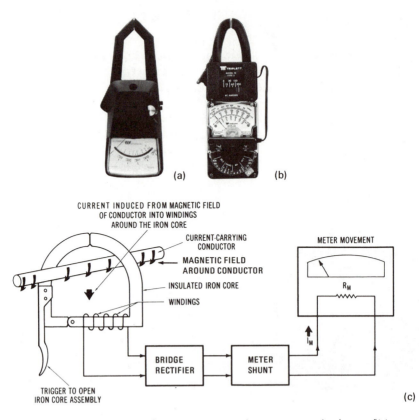

Figure 13-11 Clamp-on multimeters to measure voltage, current, and resistance. [(a), Courtesy of TIF Instruments, Inc.; (b), courtesy of Triplett Corp.]

and test instruments. The simplified circuit of a clamp-type current meter is shown in Figure 13-11(c). Current flow through a conductor creates a magnetic field around the conductor. The varying magnetic field induces a current into the iron core of the clamp portion of the meter. The meter scale is calibrated so that when a specific value of current flows in a power line, it will be indicated on the scale. Of course, the current flow in the power line is proportional to the current induced into the iron core of the clamp-type meter. Ordinarily, the clamp-type meter also has a voltage and resistance function which utilizes external test leads.

MEASUREMENT OF SPEED

A common type of electrical machine measurement is the measurement of speed. Usually, speed is measured as a rotary movement with a tachometer (see Figure 13-12). Several different principles can be used for measuring speed. One method is referred to as a dc tachometer system and is illustrated in Figure 13-13. This tachometer is connected directly to a rotating machine or piece of equipment. The principle involved

(a)

(b)

Figure 13-12 (a) Hand-held tachometer; (b) digital tachometer. (Courtesy of Biddle Instruments.)

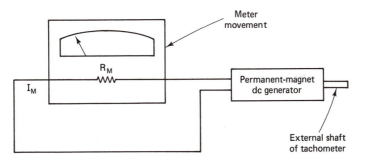

Figure 13-13 Dc tachometer system.

Figure 13-14 Phototachometer. (Courtesy of Biddle Instruments)

in this system is that as the shaft of the small permanent-magnet dc generator rotates faster, the voltage output increases in proportion to the speed of rotation. Voltage output increases can then be translated into speed changes and indicated on a calibrated tachometer scale.

Electronic tachometers, such as the photoelectric tachometer shown in Figure 13-14, are now used extensively due to their increased precision and ease of usage. In the photoelectric tachometer, movement is measured by providing a reflective material on the surface of the equipment or machine subject to measurement. The tachometer has a light source that is interrupted by the passage of the reflective material. A photocell converts the changes of light energy into electrical impulses. The electrical impulses control the movement of an indicator calibrated in revolutions per minute (r/min), or the pulse rate of a digital counter that provides a direct numerical readout of revolutions per minute.

MEASUREMENT OF VIBRATION

Vibration is a factor that must be controlled in electrical machines. In order to measure vibration in a piece of equipment, a vibration meter with a special transducer can be used. The reduction of vibration will improve the operating characteristics of machinery. The transducer used to measure vibration has an output that is proportional to the amplitude of the vibrations, sensed while a machine is in operation.

MOTOR TESTING

Electric motors convert electrical energy into mechanical energy. The *efficiency* of this energy conversion is critical in terms of motor operation. Recall that efficiency is calculated as the ratio of power output to power input of a machine. For electric motors,

$$\text{efficiency} = \frac{\text{power output (hp)}}{\text{power input (W)}} \quad \text{or} \quad \frac{\text{hp} \times 746}{\text{W}}$$

The horsepower of a motor is determined by the torque and speed of the machine. Torque may be determined by measurement with a *prony brake* assembly. A representation of a prony brake is shown in Figure 13-15. Tightening the bolts onto the special pulley causes the brake unit to turn with the motor and apply force to the suspended scale. The motor's torque is calculated as the product of the scale reading (pounds, ounces, or grams) and the measured distance shown (feet, inches, or centimeters). The prony brake method is a friction-type brake which would be impractical for large motors.

The principle of the prony brake is used for commercial units called *dynamometers,* which are used extensively for motor testing. Dynamometer testing has become very important with increased awareness of energy consumption and rising power costs. A dynamometer is used to load a motor and measure its power output directly. A load cell is commonly used to provide an accurate measure of force exerted by a motor on the dynamometer.

Measurement of input power, current, and voltage of a motor may be accomplished by using a wattmeter, ammeter, and voltmeter as shown in Figure 13-16.

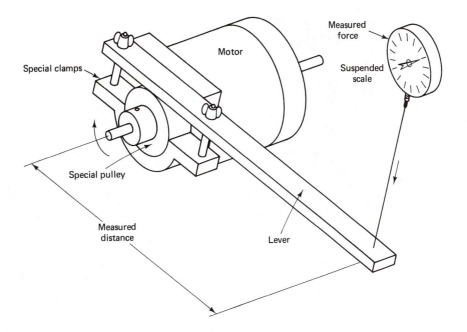

Figure 13-15 Representation of a prony brake.

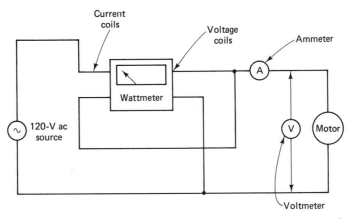

Figure 13-16 Wattmeter, ammeter, and voltmeter used to measure motor input values.

The wattmeter measures the true power converted by the motor. The power factor of a motor can be calculated as follows:

$$PF = \frac{\text{true power (W)}}{\text{apparent power (VA)}} = \frac{\text{watts}}{\text{volts} \times \text{amperes}}$$

Three-phase motor phase sequence. For three-phase motors, an important type of test is phase sequence indication. Two types of equipment used for such tests are shown in Figure 13-17. Reversed phase sequence will cause a three-phase

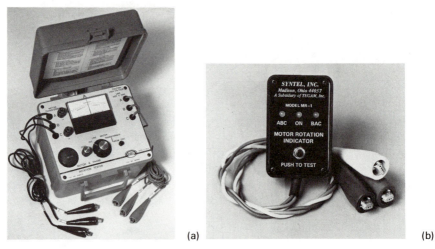

Figure 13-17 (a) Motor and phase rotation tester; (b) motor rotation indicator. [(a), Courtesy of Biddle Instruments; (b), courtesy of Tegam, Inc., Syntel Division.]

motor to rotate in the wrong direction. Correct direction of rotation often does not lend itself to trial-and-error testing. Applications such as conveyor lines, pumping systems, and interconnected drive systems require a definite direction of rotation. Thus phase sequence testing can be used to avoid problems associated with reversed direction of rotation.

REVIEW

13.1. What types of measurements may a VOM be used to perform?

13.2. What type of meter is used to measure electrical power?

13.3. What type of meter is used to measure electrical energy?

13.4. Discuss the operation of a watthour meter.

13.5. What are some methods used to measure three-phase power?

13.6. Discuss the operation of a power-factor meter.

13.7. What is a power analyzer?

13.8. How may frequency be measured?

13.9. What is the purpose of a synchroscope?

13.10. What are some applications of a megohmmeter?

13.11. Why is it often advantageous to use a clamp-on meter?

13.12. What are some methods of speed measurement?

13.13. How is machine vibration measured?

13.14. What is a prony brake?

13.15. How is a dynamometer used for motor testing?

13.16. How is the power factor of a motor determined by using a wattmeter, an ammeter, and a voltmeter?

13.17. Why is motor phase sequence important for three-phase systems?

APPENDIX 1

Metric Conversions

Appendix 1 lists some typical methods of converting quantities in the English system into values in the metric system (and vice versa).

Temperature conversions require specific procedures. These procedures follow:

1. To convert degrees Fahrenheit to degrees Celsius, use one of the following formulas:

$$^\circ C = \frac{5}{9}(^\circ F - 32^\circ) \qquad \text{or} \qquad ^\circ C = \frac{^\circ F - 32^\circ}{1.8}$$

2. To change degrees Celsius to degrees Fahrenheit, use one of the following formulas:

$$^\circ F = \frac{9}{5}(^\circ C) + 32^\circ \qquad \text{or} \qquad ^\circ F = 1.8(^\circ C) + 32^\circ$$

3. Degrees Celsius may be converted to Kelvin by using

$$^\circ C = K - 273.16^\circ$$

4. Kelvin can be converted to degrees Celsius by simply changing the formula in step 3 to

$$K = ^\circ C + 273.16^\circ$$

5. If you wish to change degrees Fahrenheit to degrees Rankine, use this formula:

$$^\circ F = ^\circ R - 459.7^\circ$$

6. Then, by changing the preceding formula, degrees Rankine can be converted to degrees Fahrenheit.

$$^\circ R = {}^\circ F + 459.7^\circ$$

By using a combination of the formulas listed above, temperature values given for any scale can be changed to a temperature on one of the three other scales. Figure A1-1 gives a comparison of the various temperature scales.

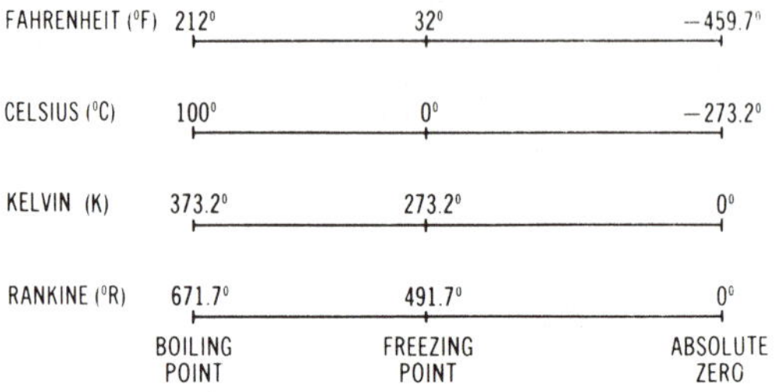

FAHRENHEIT (°F)	212°	32°	−459.7°
CELSIUS (°C)	100°	0°	−273.2°
KELVIN (K)	373.2°	273.2°	0°
RANKINE (°R)	671.7°	491.7°	0°
	BOILING POINT	FREEZING POINT	ABSOLUTE ZERO

Figure A1-1   Comparison of temperature scales.

Tables A1-1 through A1-6 list the multiplying factors for conversion of units for length, area, mass, volume, cubic measure, and electrical units.

TABLE A1-1   LENGTH CONVERSIONS

Known Quantity	Multiply by	Quantity To Find
inches (in)	2.54	centimeters (cm)
feet (ft)	30	centimeters (cm)
yards (yd)	0.9	meters (m)
miles (mi)	1.6	kilometers (km)
millimeters (mm)	0.04	inches (in)
centimeters (cm)	0.4	inches (in)
meters (m)	3.3	feet (ft)
meters (m)	1.1	yards (yd)
kilometers (km)	0.6	miles (mi)
centimeters (cm)	10	millimeters (mm)
decimeters (dm)	10	centimeters (cm)
decimeters (dm)	100	millimeters (mm)
meters (m)	10	decimeters (dm)
meters (m)	1000	millimeters (mm)
dekameters (dam)	10	meters (m)
hectometers (hm)	10	dekameters (dam)
hectometers (hm)	100	meters (m)
kilometers (km)	10	hectometers (hm)
kilometers (km)	1000	meters (m)

TABLE A1-2   AREA CONVERSIONS

Known Quantity	Multiply by	Quantity To Find
square inches (in^2)	6.5	square centimeters (cm^2)
square feet (ft^2)	0.09	square meters (m^2)
square yards (yd^2)	0.8	square meters (m^2)
square miles (mi^2)	2.6	square kilometers (km^2)
acres	0.4	hectares (ha)
square centimeters (cm^2)	0.16	square inches (in^2)
square meters (m^2)	1.2	square yards (yd^2)
square kilometers (km^2)	0.4	square miles (mi^2)
hectares (ha)	2.5	acres
square centimeters (cm^2)	100	square millimeters (mm^2)
square meters (m^2)	10,000	square centimeters (cm^2)
square meters (m^2)	1,000,000	square millimeters (mm^2)
ares (a)	100	square meters (m^2)
hectares (ha)	100	ares (a)
hectares (ha)	10,000	square meters (m^2)
square kilometers (km^2)	100	hectares (ha)
square kilometers (km^2)	1,000,000	square meters (m^2)

TABLE A1-3   MASS CONVERSIONS

Known Quantity	Multiply by	Quantity To Find
ounces (oz)	28	grams (g)
pounds (lb)	0.45	kilograms (kg)
tons	0.9	tonnes (t)
grams (g)	0.035	ounces (oz)
kilograms (kg)	2.2	pounds (lb)
tonnes (t)	100	kilograms (kg)
tonnes (t)	1.1	tons
centigrams (cg)	10	milligrams (mg)
decigrams (dg)	10	centigrams (cg)
decigrams (dg)	100	milligrams (mg)
grams (g)	10	decigrams (dg)
grams (g)	1000	milligrams (mg)
dekagram (dag)	10	grams (g)
hectogram (hg)	10	dekagrams (dag)
hectogram (hg)	100	grams (g)
kilograms (kg)	10	hectograms (hg)
kilograms (kg)	1000	grams (g)
metric tons (t)	1000	kilograms (kg)

TABLE A1-4　VOLUME CONVERSIONS

Known Quantity	Multiply by	Quantity To Find
milliliters (mL)	0.03	fluid ounces (fl oz)
liters (L)	2.1	pints (pt)
liters (L)	1.06	quarts (qt)
liters (L)	0.26	gallons (gal)
gallons (gal)	3.8	liters (L)
quarts (qt)	0.95	liters (L)
pints (pt)	0.47	liters (L)
cups (c)	0.24	liters (L)
fluid ounces (fl oz)	30	milliliters (mL)
teaspoons (tsp)	5	milliliters (mL)
tablespoons (tbsp)	15	milliliters (mL)
liters (L)	1000	milliliters (mL)

TABLE A1-5　CUBIC MEASURE CONVERSIONS

Known Quantity	Multiply by	Quantity To Find
cubic meters (m^3)	35	cubic feet (ft^3)
cubic meters (m^3)	1.3	cubic yards (yd^3)
cubic yards (yd^3)	0.76	cubic meters (m^3)
cubic feet (ft^3)	0.028	cubic meters (m^3)
cubic centimeters (cm^3)	1000	cubic millimeters (mm^3)
cubic decimeters (dm^3)	1000	cubic centimeters (cm^3)
cubic decimeters (dm^3)	1,000,000	cubic millimeters (mm^3)
cubic meters (m^3)	1000	cubic decimeters (dm^3)
cubic meters (m^3)	1	steres
cubic feet (ft^3)	1728	cubic inches (in^3)
cubic feet (ft^3)	28.32	liters (L)
cubic inches (in^3)	16.39	cubic centimeters (cm^3)
cubic meters (m^3)	264	gallons (gal)
cubic yards (yd^3)	27	cubic feet (ft^3)
cubic yards (yd^3)	202	gallons (gal)
gallons (gal)	231	cubic inches (in^3)

TABLE A1-6　ELECTRICAL UNIT CONVERSIONS

Known Quantity	Multiply by	Quantity To Find
Btu per minute	0.024	horsepower (hp)
Btu per minute	17.57	watts (W)
horsepower (hp)	33,000	foot-pounds per min (ft-lb/min)
horsepower (hp)	746	watts (W)
kilowatts (kW)	57	Btu per minute
kilowatts (kW)	1.34	horsepower (hp)
watts (W)	44.3	foot-pounds per min (ft-lb/min)

APPENDIX 2

Review of Trigonometry and Right Triangles

Trigonometry is a very valuable form of mathematics for anyone who studies electricity/electronics. Trigonometry deals with angles and triangles, particularly the right triangle, which has one angle of 90°. An example of a right triangle is shown in Figure A2-1. This example illustrates how resistance, reactance, and impedance are related in ac circuits. Resistance (R) and reactance (X) are quantities that are 90° apart, forming a right angle. The law of right triangles, known as the *Pythagorean theorem,* may be used to solve for the value of any right-triangle side. This theorem states that in any right triangle, the square of the hypotenuse is equal to the sum of the squares of the other two sides. With reference to Figure A2-1, the Pythagorean theorem is expressed mathematically as

$$Z^2 = R^2 + X^2$$

or

$$Z = \sqrt{R^2 + X^2}$$

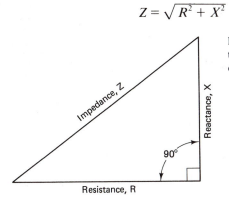

Figure A2-1 Right triangle which shows the relationship of R, X, and Z in an ac circuit.

By using trigonometric relationships, problems dealing with phase angles, power factor, and reactive power in ac circuits can be solved. The three most common trigonometric functions are the *sine,* the *cosine,* and the *tangent.* These functions are the ratios of the sides of the triangle, which determine the size of the angles. Figure A2-2 illustrates how these ratios are expressed mathematically. Their values can be found by using a calculator or in the table of natural trigonometric ratios shown in Table A2-1. This table is for angles from 0 to 90°. To find these values, use the headings at the top of the table and the angle listings from top to bottom in the left-hand column.

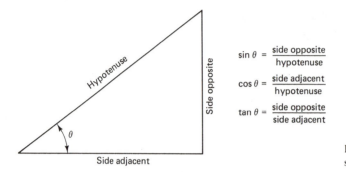

$$\sin \theta = \frac{\text{side opposite}}{\text{hypotenuse}}$$

$$\cos \theta = \frac{\text{side adjacent}}{\text{hypotenuse}}$$

$$\tan \theta = \frac{\text{side opposite}}{\text{side adjacent}}$$

Figure A2-2 Trigonometric ratios of the sides of a right triangle.

This process can be reversed to find the size of an angle when the ratios of the sides are known. The term "arc function" or "inverse function" is used to indicate this process. For example, the notation "inv sine $0.9455 = \theta$" means that θ is the angle whose sine function is x. Thus, if "inv sine $0.9455 = \theta$," look through the listing of sine functions in Figure A2-3 and find that angle θ is 71° or use the inverse sine function on a calculator.

Figure A2-3 Trigonometric ratios for second, third, and fourth quadrant angles.

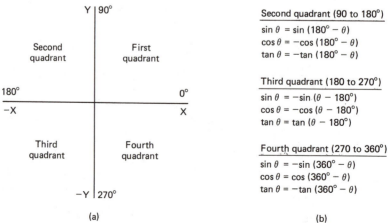

Second quadrant (90 to 180°)

$\sin \theta = \sin (180° - \theta)$
$\cos \theta = -\cos (180° - \theta)$
$\tan \theta = -\tan (180° - \theta)$

Third quadrant (180 to 270°)

$\sin \theta = -\sin (\theta - 180°)$
$\cos \theta = -\cos (\theta - 180°)$
$\tan \theta = \tan (\theta - 180°)$

Fourth quadrant (270 to 360°)

$\sin \theta = -\sin (360° - \theta)$
$\cos \theta = \cos (360° - \theta)$
$\tan \theta = -\tan (360° - \theta)$

(a)   (b)

TABLE A2-1 TABLE OF NATURAL TRIGONOMETRIC RATIOS

Sines, Cosines, and Tangents of Angles from 1 to 90°

Angle	Sine	Cosine	Tangent	Angle	Sine	Cosine	Tangent	Angle	Sine	Cosine	Tangent
1°	0.0175	0.9998	0.0175	31°	0.5150	0.8572	0.6009	61°	0.8746	0.4848	1.8040
2°	0.0349	0.9994	0.0349	32°	0.5299	0.8480	0.6249	62°	0.8829	0.4695	1.8807
3°	0.0523	0.9986	0.0524	33°	0.5446	0.8387	0.6494	63°	0.8910	0.4540	1.9626
4°	0.0698	0.9976	0.0699	34°	0.5592	0.8290	0.6745	64°	0.8988	0.4384	2.0503
5°	0.0872	0.9962	0.0875	35°	0.5736	0.8192	0.7002	65°	0.9063	0.4226	2.1445
6°	0.1045	0.9945	0.1051	36°	0.5878	0.8090	0.7265	66°	0.9135	0.4067	2.2460
7°	0.1219	0.9925	0.1228	37°	0.6018	0.7986	0.7536	67°	0.9205	0.3907	2.3559
8°	0.1392	0.9903	0.1405	38°	0.6157	0.7880	0.7813	68°	0.9272	0.3746	2.4751
9°	0.1564	0.9877	0.1584	39°	0.6293	0.7771	0.8098	69°	0.9336	0.3584	2.6051
10°	0.1736	0.9848	0.1763	40°	0.6428	0.7660	0.8391	70°	0.9397	0.3420	2.7475
11°	0.1908	0.9816	0.1944	41°	0.6561	0.7547	0.8693	71°	0.9455	0.3256	2.9042
12°	0.2079	0.9781	0.2126	42°	0.6691	0.7431	0.9004	72°	0.9511	0.3090	3.0777
13°	0.2250	0.9744	0.2309	43°	0.6820	0.7314	0.9325	73°	0.9563	0.2924	3.2709
14°	0.2419	0.9703	0.2493	44°	0.6947	0.7193	0.9657	74°	0.9613	0.2756	3.4874
15°	0.2588	0.9659	0.2679	45°	0.7071	0.7071	1.0000	75°	0.9659	0.2588	3.7321
16°	0.2756	0.9613	0.2867	46°	0.7193	0.6947	1.0355	76°	0.9703	0.2419	4.0108
17°	0.2924	0.9563	0.3057	47°	0.7314	0.6820	1.0724	77°	0.9744	0.2250	4.3315
18°	0.3090	0.9511	0.3249	48°	0.7431	0.6691	1.1106	78°	0.9781	0.2079	4.7046
19°	0.3256	0.9455	0.3443	49°	0.7547	0.6561	1.1504	79°	0.9816	0.1908	5.1446
20°	0.3420	0.9397	0.3640	50°	0.7660	0.6428	1.1918	80°	0.9848	0.1736	5.6713
21°	0.3584	0.9336	0.3839	51°	0.7771	0.6293	1.2349	81°	0.9877	0.1564	6.3138
22°	0.3746	0.9272	0.4040	52°	0.7880	0.6157	1.2799	82°	0.9903	0.1392	7.1154
23°	0.3907	0.9205	0.4245	53°	0.7986	0.6018	1.3270	83°	0.9925	0.1219	8.1443
24°	0.4067	0.9135	0.4452	54°	0.8090	0.5878	1.3763	84°	0.9945	0.1045	9.5144
25°	0.4226	0.9063	0.4663	55°	0.8192	0.5736	1.4281	85°	0.9962	0.0872	11.4301
26°	0.4384	0.8988	0.4877	56°	0.8290	0.5592	1.4826	86°	0.9975	0.0698	14.3007
27°	0.4540	0.8910	0.5095	57°	0.8387	0.5446	1.5399	87°	0.9986	0.0523	19.0811
28°	0.4695	0.8829	0.5317	58°	0.8480	0.5299	1.6003	88°	0.9994	0.0349	28.6363
29°	0.4848	0.8746	0.5543	59°	0.8572	0.5150	1.6643	89°	0.9998	0.0174	57.2900
30°	0.5000	0.8660	0.5774	60°	0.8660	0.5000	1.7321	90°	1.0000	0.0000	

Trigonometric ratios hold true for angles of any size; however, angles in the first quadrant of a standard graph (0 to 90°) are used as a reference. To solve for angles greater than 90° (second-, third-, and fourth-quadrant angles), they must be converted to first-quadrant angles (see Figure A2-3). All first-quadrant angles are positive functions, while for angles in the second, third, and fourth quadrants there are two negative functions and one positive function in each.

Index